QUESTIONS OF TIME AND TENSE

QUESTIONS OF TIME AND TENSE

edited by

ROBIN LE POIDEVIN

CLARENDON PRESS · OXFORD
1998

Oxford University Press, Great Clarendon Street, Oxford OX2 6DP

Oxford New York

Athens Auckland Bangkok Bogota Bombay Buenos Aires
Calcutta Cape Town Dar es Salaam Delhi Florence Hong Kong Istanbul
Karachi Kuala Lumpur Madras Madrid Melbourne Mexico City
Nairobi Paris Singapore Taipei Tokyo Toronto Warsaw

and associated companies in
Berlin Ibadan

Oxford is a trade mark of Oxford University Press

Published in the United States
by Oxford University Press Inc., New York

British Library Cataloguing in Publication Data
Data available

Library of Congress Cataloging in Publication Data
Data available

ISBN 0–19–823695–6

1 3 5 7 9 10 8 6 4 2

Typeset by Graphicraft Typesetters Ltd, Hong Kong
Printed in Great Britain
on acid-free paper by
Bookcraft (Bath) Ltd,
Midsomer Norton, Somerset

In memory of
Murray MacBeath, 1946–1995

PREFACE

WITH one exception, all the papers appearing here were written for this volume. The exception is Jeremy Butterfield's 'Seeing the Present', which fitted in so well with the theme of the collection that it seemed entirely appropriate to include it.

It is very sad to record that, during the preparation of *Questions of Time and Tense*, Murray MacBeath, who was going to contribute, died at the age of 48. His original and lively writings on time have done much to make the subject accessible. This collection is dedicated to his memory.

CONTENTS

Notes on the Contributors   xi

Introduction   1

1. The Past, Present, and Future of the Debate about Tense   13
   Robin Le Poidevin

2. Tense and Persistence   43
   E. J. Lowe

3. Seeing the Present   61
   Jeremy Butterfield

4. Tense and Emotion   77
   David Cockburn

5. Real Times and Possible Worlds   93
   Heather Dyke

6. Time as Spacetime   119
   Graham Nerlich

7. Absolute Simultaneity and the Infinity of Time   135
   Quentin Smith

8. Freedom and the New Theory of Time   185
   L. Nathan Oaklander

9. Morality, the Unborn, and the Open Future   207
   Piers Benn

10. The Tensed vs. Tenseless Theory of Time: A Watershed
    for the Conception of Divine Eternity   221
    William Lane Craig

11. Time and Trinity   251
    Paul Helm

12. Tense and Egocentricity in Fiction   265
    Gregory Currie

Bibliography   285

Index of Names   291

NOTES ON THE CONTRIBUTORS

PIERS BENN is Lecturer in Philosophy at the University of Leeds. He is the author of *Ethics* (1997) and articles on normative ethics.

JEREMY BUTTERFIELD is Senior Research Fellow of All Souls College, Oxford, and a Fellow of the British Academy. He is the editor of *Language, Mind and Logic* (1986) and (with others) of *Spacetime* (1996). He has published numerous articles on the philosophy of quantum mechanics.

DAVID COCKBURN is Senior Lecturer in Philosophy at the University of Wales, Lampeter. He is the author of *Other Human Beings* (1990) and *Other Times: Philosophical Perspectives on Past, Present and Future* (1997), and editor of *Human Beings* (1991). He is currently working on the problem of causation.

WILLIAM LANE CRAIG is Research Professor of Philosophy at the Talbot School of Theology. He is the author of *The Kalam Cosmological Argument* (1979) and *Divine Foreknowledge and Human Freedom* (1992), and, with Quentin Smith, *Theism, Atheism and Big Bang Cosmology* (1993). He is currently completing a two-volume study of divine eternity entitled *God, Time and Eternity*.

GREGORY CURRIE is Professor of Philosophy and Head of the School of Arts at Flinders University, Adelaide, and a fellow of the Australian Academy of the Humanities. He is the author of *An Ontology of Art* (1989), *The Nature of Fiction* (1990), *Image and Mind: Film, Philosophy and Cognitive Science* (1995), and of articles in *Mind*, *American Philosophical Quarterly*, and other philosophical journals. He is currently working on a theory of the imagination.

HEATHER DYKE is Lecturer in Philosophy at the University of Otago. She took her Ph.D. at the University of Leeds. Her dissertation was a defence of the token-reflexive version of the new tenseless theory of time.

PAUL HELM is Professor of the History and Philosophy of Religion at King's College London. He is the author of *Eternal God* (1988), *The Providence of God* (1993), and *Belief Policies* (1994). His research interests include all aspects of the philosophy of religion, and the history of Calvinism.

ROBIN LE POIDEVIN is Senior Lecturer and Head of the School of Philosophy at the University of Leeds. He is the author of *Change, Cause and*

Contradiction (1991) and *Arguing for Atheism* (1996), and co-editor, with Murray MacBeath, of *The Philosophy of Time* (1993). At present he is writing on temporal representation.

E. J. LOWE is Professor of Philosophy at the University of Durham. He is the author of *Kinds of Being: A Study of Individuation, Identity and the Logic of Sortal Terms* (1989), *Locke on Human Understanding* (1995), and *Subjects of Experience* (1996). He is currently working on a book entitled *Substance, Identity and Time*.

GRAHAM NERLICH is Emeritus Professor of Philosophy at the University of Adelaide. He is the author of *The Shape of Space* (1976, 2nd edition 1994), *Values and Valuing* (1989), and *What Spacetime Explains* (1994). He is currently working on the question of how the emotions might bear on the foundations of value.

L. NATHAN OAKLANDER is Professor of Philosophy at the University of Michigan-Flint. His main publications are *Temporal Relations and Temporal Becoming* (1984), *The New Theory of Time* (co-edited with Quentin Smith, 1994), and *Existentialist Philosophy: An Introduction* (1996). At present he is working on a book on time, freedom, and the self.

QUENTIN SMITH is Professor of Philosophy at Western Michigan University. He is the author of *The Felt Meanings of the World* (1986), *Language and Time* (1993), *Theism, Atheism and Big Bang Cosmology* (co-authored with William Lane Craig, 1993), *Time, Change and Freedom* (co-authored with L. Nathan Oaklander, 1995), and *The Question of Ethical and Religious Meanings* (1997). He is currently completing *The Uncreated Universe*, co-authored with Adolf Grünbaum.

Introduction

One of the central and most rapidly developing debates in contemporary metaphysics concerns the status of our ordinary division of time into past, present, and future. On one side of the debate stand the *tensed theorists*, who take seriously our intuitive conception of time, expressed metaphorically (or perhaps not so metaphorically) by the picture of time 'flowing'. On the other side stand the *tenseless theorists*, who deny the reality of temporal passage, and take our intuitive conception simply to reflect our perspective on time rather than the nature of time itself.

In recent years, the debate has centred around two issues:

1. What, in reality, makes true our various tensed judgements?
2. Are future events as real as present events?

Let us look at these in order. Examples of tensed judgements are: 'The party will take place tomorrow,' 'The war was a very long time ago,' 'It is now snowing in Aberdeen.' According to the tensed theory, what makes these judgements true are *tensed facts*, the fact, for example, that snowing in Aberdeen is, in some non-perspectival sense, present. This is a transient property of events, so that what is now future will become past with the passage of time. The truth-makers of tensed judgements, then, according to this theory, obtain at some times but not others. The tenseless theory, in contrast, rejects the existence of tensed facts, and explains the truth of tensed judgements in terms of unchanging temporal facts. On one version of the tenseless theory, a token present-tensed judgement, 'It is now snowing in Aberdeen,' is true if and only if the token is simultaneous with the event it describes. If such truth-conditions obtain, they obtain for all times, so tensed tokens have a time-invariant truth-value.

Turning to the second question, the issue is sometimes put as follows: are present statements about the future determinately true or false? Defenders of the unreality of the future may assert one of the following: (i) statements about the future have no truth-value, since there are no future facts to make them true or false; or (ii) statements about the future only have truth-value to the extent that present states of affairs determine the future. How does the tensed/tenseless debate as characterized above map onto the one about the reality of the future? The tenseless stance is quite clear: all times are equally real, so there are truth-makers for future-tense statements which consequently have determinate truth-values. The tensed stance is not so obvious. On the face of it, it would be possible to take a tensed view of the truth-conditions of tensed judgements while asserting the equal reality of all times. However, the plausibility of the view that the future is unreal or indeterminate, and its apparent links with our freedom of action, makes it a popular argument in favour of the tensed theory. It has also been argued that tensed theory is actually committed to the unreality, not just of the future, but of the past also.

Much has been written on these two questions, and in defence of the tensed or tenseless theories, over the last twenty years. More recently, the debate has intensified as a result of new formulations both of the tensed theory and of the tenseless theory. My characterization of the debates above would no longer be regarded as uncontentious.[1] Some tensed theorists, for example, would object to my saddling them with the idea that presentness is a property of events. But, whatever the details of formulation, the outcome of the debate is highly significant, as questions of time are intimately connected with debates in other areas of philosophy. The purpose of this volume is to make explicit those connections, to show how the debate between the two theories of time can illuminate other issues in metaphysics, philosophy of mind, philosophy of language, philosophy of science, philosophy of religion, ethics, and aesthetics, and, conversely, to show how debates in other areas of philosophy have a bearing on the debate about tense. In this Introduction I shall briefly sketch these connections, leaving a more detailed study of the development of the two theories for the first essay. For convenience, I shall group the points of contact between time and other problems under seven headings.

1. Diachronic identity

A central issue in metaphysics concerns the persistence, or identity through time, of objects. A long-standing philosophical tradition has attached significance to the commonsense distinction between terms such as 'statue' or 'chair', on the one hand, and 'concert' or 'war' on the other. According to this tradition the first kind of term is used to name continuants, whereas the second kind is used to name events or processes. The supposed difference is that

---

[1] The tensed and tenseless theories have a number of variants. On one version of the tensed theory (described as the 'tensed date theory' in Ch. 1 of this volume), the truth-value of a tensed token does *not* vary over time. Most tensed theorists treat presentness as non-relational. An exception is Tooley (1997). His position is summarized in Ch. 1 below.

continuants can be said to change, and therefore persist through change, whereas events do not, having instead different *temporal parts* at different times. However, the tenseless theory of time appears to be at odds with this distinction, providing grounds for the idea that ordinary objects such as chairs have temporal parts, and therefore cannot be said to be continuants in the traditional sense. The connection here can be presented rather informally as follows. If time flows, then objects within time change their temporal position. For example, something which existed during 1996 may, when 1996 becomes past, come to exist in 1997. Objects can in this sense be said to persist through time.[2] However, if, as the tenseless theory holds, there is no flow of time, then objects cannot shift their temporal position. What exists at one time is so located from any temporal perspective. Consequently what exists at one time cannot be strictly identical with anything existing at any other time.

Some tenseless theorists (though not all: see e.g. Mellor 1981) have accepted this, and take it to constitute a revisionist metaphysics. Some tensed theorists, conversely, have used the point as an argument against tenseless theory. But whether or not the tenseless theory does have this ontological consequence has been a matter for recent debate. The issue is addressed in the essay by Lowe, who argues that the tenseless theory is indeed committed to a temporal part ontology. This now raises further issues, since if Lowe is right the tenseless theorist faces the challenge of how to account for change. For if change is a matter of the *same thing* having different properties at different times, then it would appear that any denial of the existence of continuants entails a denial of change.

On the other hand, the defender of temporal parts can argue that the notion of strict diachronic identity is, quite independently of questions concerning the nature of time, a deeply problematic one. Since Derek Parfit's influential 1971 paper 'Personal Identity' writings on identity have often focused on problem cases where, for example, one object 'becomes' two, as in the transplantation of each of the hemispheres of a subject's brain into a different body. If we are prepared to make a few empirical assumptions—e.g. about the possibility of the two hemispheres being functionally isomorphic—we seem to be faced with a case where each of two subjects is psychologically continuous (and physically continuous, if we take the brain to be the only relevant

---

[2] I am grateful to a reader for Oxford University Press for pointing out to me that there is another way in which an item can change position in time, one which does not imply persistence. An event can become increasingly past, but we do not thereby wish to count events as continuants. In terms of the A-series/B-series distinction (defined in Ch. 1), events and objects which have ceased to exist change their A-series position. Persisting objects, in contrast, change their present B-series position.

part of the body where questions of personal identity are concerned) with the original subject. The notion of strict diachronic identity is simply not equipped to deal with such cases, it seems. If, as Parfit and others have argued, we should therefore abandon the notion, tenseless theorists are well advised to adopt temporal part theory. Their view of the nature of time would then simply be another route to a correct view of identity.

2. *The experience of time*

Tense is a pervasive aspect of our experience of the world: we cannot help but view things from the perspective of the present moment, we have no direct experience of the future, and there is a powerful sense of the passage of time. Moreover, our representation of ourselves as persisting persons favours the continuant ontology which many have seen to be essentially tied to tensed theory. Because of these features of our experience, tenseless theorists have recently tried to give an account of our perception of the world that is consistent with their account of the reality of time. In doing this, the tenseless theorist must avoid a trap analogous to the one which Dummett thinks McTaggart (who denies the reality of time) falls into:

[The view that time is unreal] seems self-refuting in something of the way in which, as McTaggart himself points out, the view that evil is an illusion is self-refuting: that is, if there is no evil, the illusion that there is evil is certainly evil. . . . Clearly, even if the world is really static, our apprehension of it changes. It does not help to say that we are even mistaken about what we think we see, because the fact would remain that we still make different such mistakes at different times. (Dummett 1960: 356)

To put it another way: if time is mind-dependent, then there is something which is genuinely temporal, namely minds. A similar worry haunts the denial of tense: if tense is mind-dependent, then minds at least must themselves be tensed. That is, the perception of an otherwise tenseless world, to generate the illusion of the passage of time, must itself be subject to that passage. This is certainly an untenable position, since minds are part of the world, and if minds are tensed the world itself cannot be tenseless. So when the tenseless theorist says that tense is mind-dependent—i.e. that there are tensed thoughts, but no tensed facts—he does *not* mean that tense attaches to thoughts in exactly the same way as the tensed theorist thinks it attaches to facts. (Consider this analogy: when we say beauty is in the eye of the beholder we do not mean that the eye itself is beautiful.) The explanation of our experience of time, then, must for the tenseless theorist involve nothing which can only be articulated in terms of past, present, and future.

Butterfield's essay attempts, among other things, to reconcile tenseless theory with an aspect of experience which is perhaps the one most responsible for our belief in the passage of time: the fact that we appear to share a common now, but not a common here. I share a now with those I am in communication with to the extent that we largely agree on what is now the case. This, suggests Butterfield, is explicable by a contingent fact which has nothing to do with the passage of time: the fact that the rate of change of most observed objects is far slower than the rate of communication between individuals, and slower than the rate at which we directly receive information about those objects (through reflected light hitting the retina, for example). The consequence of this is that my current beliefs about what state objects are in, based on both my observations and communication with other observers, are likely to be up to date.

The essay by Cockburn, in contrast, is critical of the idea that the tenseless theorist can take a conciliatory attitude towards our experience of time. Cockburn focuses on the emotions, and argues that the tenseless theory is obliged to be revisionary. A crucial problem here is one noted by Arthur Prior. We anticipate future suffering with dread and feel relief when it is over. The connection between belief and emotion is surely more than causal: we do not feel relief simply as a causal consequence of our belief that suffering is over. The belief serves as a justification for the emotion. It is appropriate to feel relief. Now, as Prior (1959) pointed out, if we replace the tensed belief by a supposedly equivalent tenseless one, so that, for example, instead of 'Suffering is now over,' we have 'Suffering is earlier than this thought,' we no longer have a belief which makes relief any more appropriate than dread, or indifference. Yet, according to the tenseless theorist, the tensed belief is made true by a purely tenseless fact, such as that expressed by the second belief. Cockburn argues that the tenseless theorist is faced with a dilemma: either we should revise our emotional attitudes to events, so that (as Spinoza argued) we should regard an event in the same way whether it is past, present, or future, or we should abandon our understanding of emotions as states which can be rationalized.

3. The language of time

The original version of the tenseless theory (as espoused, for example, by Russell) attempted to capture the meaning of tensed statements in tenseless terms. Thus, 'It is now snowing' would have been regarded as translatable, without loss of meaning, into some such sentence as 'Snowing is simultaneous with this judgement.' However, proponents of what has come to be called the *new tenseless theory*, developed since 1980, have conceded that such a translation is impossible, and that tensed meaning is irreducible. Instead, they attempt to provide the *truth-conditions* of tensed statements in tenseless terms. Typically, 'now' and other tensed terms are treated as indexicals. In one version of the theory, a token of 'It is now snowing' is taken to be true if and only if the token occurs simultaneously with an event of its snowing. In another, the token is true if and only if the event occurs at the time the token occurs. These two versions may at first sight seem to be two ways of saying exactly the same thing. But, as we shall see in the first essay of this collection, the two are not logically equivalent.

The language of time bears a number of analogies to the language of alethic modality—talk of the possibly true and the necessarily true. For example, just as we can represent past- and future-tensed sentences as locating states of affairs at other times, so we may represent modal sentences, such as 'Wilson might have won the 1970 British general election,' as locating states of affairs in other possible worlds. Further, an anti-realist treatment of past and future ('only what is present is real') mirrors an anti-realist treatment of other worlds ('only what is actual is real; other worlds are just consistent stories about how this world might have been'). But how significant are these analogies? It would seem perfectly coherent to adopt an anti-realist treatment of other worlds whilst embracing realism about other times. However, the connections may be more than skin-deep. Tenseless theorists such as Hugh Mellor have argued against the existence of tensed facts from the starting-point of a token-reflexive treatment of tensed terms such as 'now' (the details of the argument are outlined in Le Poidevin's essay). That is, if we take the truth-value of a given token of, for example, 'The maniac is now approaching' to depend on the time at which the token occurs, we are, so the argument goes, committed to treating all times as equally real. Now the term 'the actual world' is plausibly regarded as behaving like a token-reflexive: that is, a given use of it picks out whichever world it is used in, so that the truth-value of a modally tensed judgement such as 'The maniac is actually approaching' depends on the world in which the judgement is made. But if this is so, then, by parity of reasoning, we should take other worlds to be as real as the actual world. Does the argument for tenseless theory then lead ultimately to modal realism? This would be an unhappy conclusion for most tenseless theorists. Dyke argues that much more is needed to motivate a metaphysical position over modality than an indexical analysis of modal terms, and further that modal realism suffers from a circularity which does not threaten its temporal counterpart.

4. *Physics and the present*

One of the most dramatic aspects of the Special Theory of Relativity, under its standard interpretation, is its treatment of simultaneity as frame-relative. But the notion that simultaneity is essentially relative appears to be at odds with tensed theory, for if the distinction between past, present, and future is non-relational, then all present events would appear to be *absolutely* simultaneous with each other, not merely simultaneous with respect to some arbitrarily selected frame.

The two essays in this volume on the topic of what physics has to say about the nature of the present take radically opposing viewpoints. For Nerlich, physics has indeed shown that presentness is a purely relational feature of events, and that time is simply an aspect of spacetime. Smith rejects the notion that physics has superseded metaphysics where questions about time are concerned. (Perhaps the most extreme statement of this view is Hilary Putnam's (1967) suggestion that there are no longer any philosophical problems about time.) His essay is critical of the traditional dichotomy between physics and metaphysics, and argues that the theories of time which appear in the Special and General Theories of Relativity and in orthodox interpretations of quantum mechanics are false metaphysical theories, based on verificationist assumptions. Once we reject this verificationist basis, Smith argues, physics can be seen to imply that temporal relations are absolute and not relative to a reference frame, and that past and future time are infinite. Other consequences are also drawn out in Smith's essay.

Smith's own version of the tensed theory, articulated most fully in his *Language and Time* (1993), is an unusual one. He holds that presentness inheres in every state of affairs, a doctrine he names 'presentism'. This is not to be confused with the idea that only the present exists, however, for he treats all times as equally real in the sense that they are within the domain of quantification. Nerlich's essay takes issue with Smith's presentism, and argues that some propositions, at least, must ascribe purely tenseless properties to the world. He concludes by offering a sketch of an error theory about tense, i.e. a tenseless explanation of why we are led to represent the world as tensed.

5. *Time and the moral agent*

There is a familiar argument to the effect that the tenseless theory of time creates difficulties for our view of ourselves as moral agents. It goes as follows: if future-tensed statements are determinately true or false, as the tenseless theory asserts, then there is a sense in which the future is fixed and this undermines the belief that we are completely free agents and so responsible for our future actions. Another, less explored, argument has to do with the debate on temporal parts, discussed above under 'Diachronic Identity'. There it was suggested that the temporal part ontology which, according to some, is implied by the tenseless theory amounts to a denial of real change. But our freedom surely involves the ability to bring about a change in the properties of an object. Both arguments are discussed in Oaklander's chapter. In his terminology, the first draws attention to the tenseless theory's apparent *logical* threat to freedom, and the second to a *metaphysical* threat. He also discusses the apparent *causal* threat, which is raised by the connection some have seen between tenseless theory and determinism. After correcting some misunderstandings of tenseless theory by certain writers, Oaklander argues that the tenseless theory poses none of these threats to freedom.

Mirroring the issue of our freedom to perform future actions is the issue of how far we are responsible for our past actions. Once more the tenseless theory raises difficulties. For we are responsible for past actions, so a plausible line of reasoning goes, only in so far as we are identical with the performers of those actions, and this requires the continuant ontology apparently ruled out by tenseless theory. Here again the work of Parfit is relevant (see especially Parfit 1984), for, although not primarily concerned with the nature of time, he wishes to develop a sense of moral responsibility based on the notion of survival rather than strict identity. So here is an instance where tenseless theorists can learn from work at the interface between ethics and the philosophy of mind.

On the other side, the tensed theory is not without its ethical problems. For if, as one version of the tensed theory has it, the future is unreal, future (i.e. as yet unconceived) individuals are ontologically problematic. Any apparent obligations we have towards them cannot, it seems, have an object. And how can we distinguish between those future individuals who become actual (and to whom, intuitively, we have obligations) and those who remain purely potential (and to whom, intuitively, we have no obligations)? These questions are taken up by Benn, who argues that the differences between the tenseless and tensed theories in these respects are less significant than we might at first be inclined to think.

6. *God's relation to time*

The issue of God's relation to time has long been seen as problematic. Should we regard God as being within, or outside, time? If outside, how is it possible for God to have causal interactions with the universe? If within, what sense can we make of the notion that 'all times are present to God'? The tensed/tenseless dispute gives the whole range of problems in this area a new twist. For example, if, as one version of the tensed view of time has it, the future is not real, does this entail that God does not know the future? If so, can this be reconciled with his omniscience? On the tenseless theory, there are future facts, and if an omniscient being knows all the facts that there are, then God knows the future. But for some, God's knowing the future is a greater threat to human freedom than simply the abstract fact that future-tensed statements have a determinate truth-value.

One might wonder whether the existence of a timeless mind is compatible with the tensed theory of time, whether or not that mind is omniscient. Let us assume that for any true assertion it is at least possible that God should know that the assertion is true. Now consider some true assertion of the form '*E* is now occurring.' In a well-known paper, Norman Kretzmann (1966) asks whether a timeless, or immutable, God could know tensed assertions to be true. Trivially, he could only know '*E* is now occurring' to be true if he knows its truth-conditions to obtain. But only if the truth-conditions of the assertion are tenseless could a timeless God know that they obtain. That is, he could know that the assertion was made at *t*, and that *E* occurs at *t*. What he could not know is that some tensed fact obtained, for this would require him to have a tensed perspective, which is incompatible with his being timeless.

Craig looks at the notions of divine eternity and God's creation of time, and asks how one's view of these is affected by the outcome of the debate over tense. Craig examines a number of unsuccessful attempts to reconcile the traditional doctrine of divine timelessness with the idea of God's being creatively active, and argues that reconciliation is only possible if one adopts the tenseless theory of time.

Helm considers the difficulties Richard Swinburne's view of God as existing within a tensed universe poses for the doctrine of the Trinity. If the Father's creation of the Son is a causal process within tensed time, this appears to have the consequence that it is possible for the Father to exist and not the Son, which would be quite incompatible with traditional trinitarian teachings in which God necessarily coexists with the Son. In place of Swinburne's picture, Helm presents us with an account of God as atemporal and the relation between Father and Son as an atemporal relation which is at least analogous to a causal relation. That the atemporal relation in question is in fact a causal one, as Helm argues, would, of course, be disputed by theorists who hold there to be an analytic connection between causal priority and temporal priority.

7. *The time of fiction*

Narrative fictions represent fictional events as occurring in temporal order, an order we must discern in order to understand the fiction. But the difference between earlier and later fictional events is not simply that between events which are presented earlier and those which are presented later, for many fictions employ *anachronies* (e.g. flashbacks). So we cannot reduce fictional time to facts about the temporal structure of our engagement with fiction. In so far as talk about fictional events is unproblematic, talk about there being temporal relations between those events is so. But how we are to understand those relations will be informed by our stance on the tensed/tenseless dispute, and here problems do arise. The tensed theorist typically takes statements about temporal order to be reducible to statements about tense. For example, '*E* occurs simultaneously with *F*' will reduce to some such formula as 'Either *E* and *F* are both present, or both are equally distantly past or future.' However, it is not clear that we can make sense of this reduction in the context of fictional events. For, arguably, we cannot say, of a fictional event, that it is present, without implying that we are in the same time-frame as the event in question, and hence that it is really going on. But if we cannot coherently adopt a tensed perspective on events within fiction, then fictional discourse seems to provide an example of a tenseless language of before and after which is quite independent of the language of tense. And if it does provide such an example, then one important tenet of tensed theory is threatened.

Although not wishing to present a case for the tensed theory of time, Currie argues that there are no logical barriers to attributing tense to fictional events. That is, we can coherently say, of an event, that it is fictionally true that it is present. However, although such attribution of tense to fiction is not logically incoherent, there may be considerations concerning imagination which make it inappropriate to represent a fictional event as past, present, or future.

The articles of this volume provide ample illustration of what Bill Newton-Smith (1980) has aptly termed the promiscuous character of time: its tendency to couple with many other abstract issues. And of all the various aspects of time, it is its tensed aspect, the fact that we experience time in terms of a transient now, which is the most promiscuous. However, the way in which that aspect informs discussion of other philosophical issues is not always on the surface. One needs to look more deeply. That, I hope and believe, is what the following essays do.[3]

---

[3] For more discussion of the bearing of the debate between the tensed and tenseless theories of time on other philosophical debates, see Quentin Smith's 'General Introduction', in Oaklander and Smith (1994: 1–14).

REFERENCES

Dummett, M. A. E. (1960), 'A Defence of McTaggart's Proof of the Unreality of Time', repr. in *Truth and Other Enigmas* (London: Duckworth, 1978): 351–7.

Kretzmann, N. (1966), 'Omniscience and Immutability', *Journal of Philosophy*, 63: 409–21.

Mellor, D. H. (1981), *Real Time* (Cambridge: Cambridge University Press).

Newton-Smith, W. H. (1980), *The Structure of Time* (London: Routledge & Kegan Paul).

Oaklander, L. N., and Smith, Q. (1994) (eds.), *The New Theory of Time* (New Haven: Yale University Press).

Parfit, D. (1971), 'Personal Identity', *Philosophical Review*, 80: 3–27.

—— (1984), *Reasons and Persons* (Oxford: Oxford University Press).

Prior, A. N. (1959), 'Thank Goodness That's Over', *Philosophy*, 34: 12–17.

Putnam, H. (1967), 'Time and Physical Geometry', *Journal of Philosophy*, 64: 240–7.

Smith, Q. (1993), *Language and Time* (New York: Oxford University Press).

Tooley, M. (1997), *Time, Tense, and Causation* (Oxford: Clarendon Press).

I

The Past, Present, and Future of the Debate about Tense

ROBIN LE POIDEVIN

1. McTaggart

(a) The A- and B-series

The first clear articulation of the central issue about tensed distinctions is to be found in John Ellis McTaggart's article 'The Unreality of Time' (1908), and again in chapter 27 of *The Nature of Existence* (1927). Here is how he presents that issue in the later discussion:

It might be the case that the distinction of positions in time into past, present, and future, is only a constant illusion of our minds, and that the real nature of time contains only the distinctions of the *B* series—the distinction of earlier and later. In that case we should not perceive time as it really is, though we might be able to think of it as it really is. (McTaggart 1927: § 308)

As he points out, a number of earlier writers had held time, and hence the division of past, present, and future, to be unreal, but here for the first time we find the suggestion that the nature of time may be represented simply in terms of earlier and later. McTaggart goes on to reject that suggestion, but the idea is there and it was to be developed in Russell's theory of time.

Let us look more carefully at the distinction on which the issue rests. What McTaggart calls the 'A-series' is the series of positions in time which runs from the distant past to the distant future. His 'B-series' is the series of positions which is ordered by the transitive asymmetrical relation 'earlier than' (or 'later than'). To call the B-series, as he does at one point, simply 'The series of positions which runs from earlier to later' tends to obscure the distinction. For when the point from which something is judged to be earlier and later is the *present* moment, 'earlier' and 'later' become synonymous with 'past' and 'future' respectively. The crucial feature of the B-series, however, is that positions in it are unchanging: if it is true at one time that event x is earlier than event y, then it is true at all times that x is earlier than y. Positions in the A-series, in contrast, do change: what is now present was once future and will be past.

Now when McTaggart says that these are the two ways in which we distinguish positions in time, as it appears to us, he is simply articulating a common practice with no explicit philosophical overtones. But when he raises the issue of whether an A-series is essential to the reality of time, he introduces a metaphysical inquiry, not just because of the assumption of realism in talking of time as it is independently of our representation of it, but also because it involves the underlying relation between the A-series and the B-series: are they just alternative descriptions of the same feature of the world? If so, does one have priority over the other?

What for McTaggart settles the question of what is and is not essential to time is what is needed to account for *change*, where 'change' here means change in the ordinary properties of things, as when a leaf turns brown. For McTaggart, there cannot be time without change, but we can retain his insight while making room for the idea of periods of time devoid of change by insisting that, since the reality of time entails at least the logical possibility of change, whatever is a necessary condition for the possibility of change is also a necessary condition for the existence of time. McTaggart argues that the A-series is a necessary condition for change, in that it is part of the analysis of change, and so it is a necessary condition for time. Change cannot be captured in purely B-series terms, so, in so far as B-series relations are temporal relations, there cannot be a B-series unless there is also an A-series. That suggests that the A-series is somehow prior to the B-series, that we can understand the B-series in terms of the A-series, but not vice versa. Initially, McTaggart does not spell out just how we can understand the B-series in purely A-series terms (though he does have an account of how the B-series can be derived from the A-series and some other facts about the world—see section (*c*) below), but later he offers the following analysis:

The term *P* is earlier than the term *Q*, if it is ever past while *Q* is present, or present while *Q* is future. (1927: 271)

Such an analysis presupposes, of course, that 'past' is not to be understood as 'earlier than the present', and 'future' as 'later than the present', otherwise we have a circular definition. One may be worried by the occurrence of the term 'while', with its tacit quantification over B-series times. Another suggestion, made by later writers (e.g. Gale 1968), is that the B-series statement

A is earlier than *B*

may be analysed into the following disjunction of A-series statements:

A is past and *B* is present, or *A* is present and *B* is future, or *A* is past and *B* is future, or *A* is more past than *B*, or *A* is less future than *B*.

We may take these as expressions of the ontological dependence of the B-series on the A-series. Is McTaggart also advancing a thesis about meaning? It is possible to read his analysis as implying that we cannot truly assert any B-series statement unless we could also truly assert some A-series statement, since B-series expressions are translatable by A-series expressions. However, I hesitate to ascribe this further thesis about meaning to McTaggart, for two reasons. The first is that he does not talk in terms of what can truly be asserted, but what is necessary for there to be change. The second is that he makes some puzzling remarks which suggest that the reality of the A-series is actually independent of what we can truly assert. These remarks are made in the context of his discussion of a number of objections to the idea that time depends on a real A-series. This is what he says:

322   The third objection is based on the possibility that, if time were real at all, there might be in reality several real and independent time-series. The objection, if I understand it rightly, is that every time-series would be real, while the distinctions of past, present, and future would only have a meaning within each series, and would not, therefore, be taken as absolutely real. There would be, for example, many presents. Now, of course, many points of time can be present. In each time-series many points are present, but they must be present successively. And the presents of the different time-series would not be successive, since they are not in the same time. . . . So the different time-series, which are real, must be able to exist independently of the distinction between past, present, and future.

323   I cannot, however, regard this objection as valid. No doubt in such a case, no present would be the present—it would only be the present of a certain aspect of the universe. But then no time would be the time—it would only be the time of a certain aspect of the universe. It would be a real time-series, but I do not see that the present would be less real than the time.

. . . if there were any reason to suppose that there were several distinct B series, there would be no additional difficulty in supposing that there should be a distinct A series for each B series. (1927: 271)

Let us put the objection in slightly different terms. Suppose there to be another time-series, existing independently of this one, such that no time in that series bears any temporal relation to any time in our time-series. Of that time-series, we could say, 'Some events are earlier than others,' but we would not say, 'Some events are past, while others are present,' or indeed make any other unconditional A-series statement about that time-series. We could make a conditional statement such as 'Were I located in that time-series, some event would be present,' but to make an unconditional A-series statement about an event is to place oneself in a temporal relation to that event, and *ex hypothesi* there are no temporal relations between us and the events of the other time-series. Consequently, it is possible to make a B-series statement about certain events

without being in a position to make any (unconditional) A-series statement about those events. So, if this thought experiment is intelligible, B-series
expressions are not translatable by A-series expressions. Now what we would
expect McTaggart to do is to deny the intelligibility of discourse about other
time-series, as later writers have done (see Le Poidevin 1996 for a critical
discussion of their arguments). But he does not do this. So this suggests
that his conception of the relationship between the A-series and the B-series
has nothing to do with any dependence, in terms of meaning, of B-series
statements upon A-series statements. So in the following reconstruction of
his argument for the dependence of the B-series on the A-series, we should
take the conclusion (and therefore at least one of the premisses) to be intended
as a necessary, but perhaps not an analytic, truth:

   1. B-series relations are temporal relations.
   2. There cannot be temporal relations unless there is change.
   3. There cannot be change unless time in reality has an A-series ordering.
   Therefore: There cannot be B-series relations unless time in reality has
   an A-series ordering.

(b) Russell versus McTaggart

Why does McTaggart think the A-series essential to our understanding of
change? He considers Russell's analysis of change, as presented in *The
Principles of Mathematics* (1903). Russell phrases the analysis in terms of
propositions and truth-values, but we would be doing no violence to his idea
by summarizing it as follows: an object changes if and only if it has at one
time a property which it lacks at another time. Change, then, is simply the
variation of properties over time. Nothing, in Russell's analysis, requires these
times to be ordered as an A-series. The B-series is sufficient for his purpose.
McTaggart's objection to this is that change cannot consist in a conjunction
of changeless facts, and Russell's analysis of change is precisely in terms of
changeless facts. Thus, according to Russell, the change that takes place when,
for example, a poker cools down is captured by the facts that (i) it is hot at
one time, and (ii) it is not hot at a later time. But, assuming that these times
are positions in the B-series, then facts (i) and (ii) never change: they obtain
at all times. To capture fully genuine change, for McTaggart, we must analyse it in A-series terms: the poker *was* hot and *is now* not hot. The fact that
the poker is now not hot is something that changes over time.

   One point of contention between McTaggart and Russell, then, concerns
the correct analysis of change. It has to be said that, when Russell defines
change in *The Principles of Mathematics*, the definition is formally neutral

on the issue whether the times in question are B-series or A-series times. But McTaggart is quite justified in taking it to be, implicitly, a B-series formulation because elsewhere Russell makes it clear that the B-series is not only not dependent on the A-series, but is actually the basis of the A-series. Like McTaggart, he provides an explicit analysis of one in terms of the other, but this time the postulated dependence is not just an ontological one (or, as we might put it, providing statements concerning one series with truth-conditions in terms of the other) but also one of meaning. In 'On the Experience of Time' (1915), Russell proposes that the assertion that some event is occurring now is equivalent in meaning to the tenseless assertion that that event occurs simultaneously with some sense datum. So, for Russell, A-series statements can be true even if in reality there is no A-series.

We can summarize the differences between McTaggart and Russell as follows:

McTaggart
(*a*) Change cannot adequately be analysed in B-series terms only: A-series terms are essential. Thus, for an object to change is for it to be the case that it was in a certain state and is now not in that state.
(*b*) Since there cannot be time without change, and B-series relations are temporal relations, the B-series is (ontologically) dependent on the A-series.

Russell
(*a*) Change can be adequately analysed in B-series terms only. Thus, for an object to change is for it to be in a certain state at one time and not in that state at a later time.
(*b*) Since A-series expressions can be analysed in terms of B-series expressions, the A-series is dependent, both ontologically and in terms of the meaning of the corresponding expressions, on the B-series.

Here we have the clear beginnings of the debate between the tensed and tenseless theories of time. However, although McTaggart's (*a*) and (*b*) are accepted by most modern tensed theorists, we cannot ascribe a tensed view to McTaggart because, notoriously, he held that the A-series, and hence time itself, is unreal. To the extent that he denies reality to the A-series, then, McTaggart is actually in agreement with Russell. The details of his argument have been well documented in a number of commentaries (see e.g. Dummett 1960, Mellor 1981, Farmer 1990), and since my purpose here is mainly that of showing how the opposing sides in the debate have developed, I shall give no more than the briefest outline of his argument. McTaggart holds that the A-series is unreal because it contains a contradiction which arises from the following two inconsistent propositions:

[1] Past, present and future are incompatible determinations. Every event must be one or the other, but no event can be more than one.

[2] But every event has them all. If M is past, it has been present and future. (McTaggart 1927: § 329)

Defenders of the A-series have pointed out that the appearance of contradiction soon dissolves once we realize that past, present, and future are only incompatible if applied *simultaneously*, and the sense in which every event has them all is that every event occupies each A-series position *successively*. Those that deny an A-series, and side with Russell's (*a*) (but perhaps not his (*b*), for reasons we shall encounter later), reply that the only consistent way of articulating this insight is in B-series terms. Thus, we should say that M is present at *t* (where *t* is a B-series position), future at times earlier than *t*, and past at times later than *t*. Now if A-series positions are relativized to B-series positions in this way, the question arises whether or not '*M* is present at *t*' says anything more than '*M* is simultaneous with *t*'. The tenseless theorist will hold that these expressions are equivalent.

(*c*) *The C-series*

Having denied the reality of time, McTaggart surely owes us an *error theory* of our experience of time, i.e. an explanation of why we perceive reality as having an A-series ordering when in fact it does not. Although he does go some way towards providing this, it is clear from his remarks that he feels little pressure to do so. Even those who believe in the reality of time, he says, must admit that time is not as we perceive it to be. When we talk of the present of experience, the 'specious present', we talk of something with duration. But if there were a present in reality, it would have no duration: it would simply be the extensionless point dividing past from future. Thus to deny the reality of time altogether is not such a big step from the commonsense position that we already occupy with regard to our experience of time.

However, McTaggart clearly thinks that to deny that reality has *any* kind of ordering is too counter-intuitive: our experience must be right on that score, even if we are wrong in thinking that it must be a temporal ordering. His error theory, then, is this. The world contains a non-temporal linear ordering, the 'C-series'. Now, although he does not say exactly what C-series relations are, they clearly mirror our representation of B-series relations:

349   ... If we say that the events M and N are simultaneous, we say that they occupy the same position in the time-series. And there will be some truth in this, for the realities, which we perceive as the events M and N, do really occupy the same position in a series, though it is not a temporal series.

Again, if we assert that the events M, N, O are all at different times, and are in that order, we assert that they occupy different positions in the time-series, and that the position of N is between the positions of M and O. And it will be true that the realities which we see as these events will be in a series, though not in a temporal series, and that they will be in different positions in it, and that the position of the reality which we perceive as the event N will be between the positions of the realities which we perceive as the events M and O. (1927: § 349)

And, we might add, to the experience of change in the poker there corresponds the fact that the poker is in one state at one point in the C-series and not in that state at some other point in the C-series. It is not surprising, given these correlations between experience and reality, that McTaggart admits, 'Thus the C series will not be altogether unlike the time-series as conceived by Mr Russell' (ibid., § 351). Indeed, it is tempting to say that the C-series just *is* the B-series, in all but name. McTaggart, however, insists that the B-series is in fact a construction from the C-series and the A-series:

348   For if there is a C series, where the terms are connected by permanent relations and if the terms of this series appear also to form an A-series, it will follow that the terms of the C series will also appear as a B series, those which are placed first, in the direction from past to future, appearing as earlier than those whose places are further in the direction of the future.

The A-series is thus what adds the element of temporality to the relations of the C-series, as we perceive them. But if the C-series is all that there is, we need some explanation of how our experience of the A-series, and not just of the B-series, is derived from it. Here, however, we are left in the dark. It does not help that the C-series appears to have no directionality. (Note that McTaggart talks above of C-series betweenness, but not of priority.)

In drawing the lines of the debate between Russell and McTaggart, we have found the origins of the tensed and tenseless theories of time, with Russell clearly defining a tenseless position. Despite their disagreements, however, McTaggart is much closer to Russell than he is to the tensed theorists who champion his analysis of change.

(d) The reality of the future

In his discussion of McTaggart, C. D. Broad (1927) suggested a rather different view of the A-series. We should not think of there being events located in the future and destined, as it were, to become present. Rather, we should think of the future as being unreal. As events come into being, the sum total of reality—i.e. the past and present—increases. Broad thought that this conception of the A-series avoided McTaggart's contradiction. But, as subsequent commentators have shown (e.g. Mellor 1981, Le Poidevin 1991), it is possible to reconstruct the proof of the unreality of time in a way which makes no reference to events and their A-series positions.

Broad's suggestion, however, is of interest in its own right. Arguably, it is a very intuitive conception, and provides one of the strongest motivations for the tensed view of reality. Ontological differences between the past and present on the one hand and the future on the other cannot be captured in purely B-series terms, for if it is true that some object is real at t, where t is some position in the B-series, it is true at all previous times. Ascriptions of reality can only change their truth-value if truth is relative to position in the A-series. It might be objected, however, that B-series facts *can* change. For objects must be real for B-series relations to obtain between them, so that although it is now true that two presently existing events are simultaneous, it was not true until those events came into existence. But this thought requires the notion of *present* truth, which is clearly an A-series concept. So it would seem that the A-series is still needed to capture ideas concerning the unreality of other times.

We must be careful, however, to distinguish the obvious fact that we stand in certain epistemological relations to the future which we do not stand in to the past and present (we cannot perceive the future as we can the present, and our knowledge of what will be is therefore only inferential), from the fact—if it is one—that the future simply does not exist. What Broad's view involves, in more detail, is the idea that statements about the future lack truth-value since there is nothing in reality to confer any truth-value upon them. McTaggart's objection to this view (McTaggart 1927: §§ 337–9) is that it conflicts with the plausible notion that the past sometimes determines the future. To cite his example: 'If Smith has already died childless, this intrinsically determines that no future event will be a marriage of one of Smith's grand-children' (1927: § 338). If the world is entirely deterministic, then, all future-tensed statements have a determinate truth-value.

We can reconcile Broad's world-picture with determinism, however, if we cast it in terms of what facts constitute the truth-conditions of future-tensed statements. Whereas statements about the past or present are made true or false by past or present fact, statements about the future either lack a truth-value or, if they have one, it is in virtue of past or present fact, not future fact. This is what the thesis of the unreality of the future amounts to. So, if the world were deterministic then, in virtue of what is presently the case, each future-tensed statement would have a determinate truth-value. What matters, as far as questions concerning the status of the future are concerned, is not so much whether future-tensed statements have truth-values, but what kinds of fact fix those truth-values.

Another way of capturing the unreality of the future which makes no presuppositions about determinism concerns the status of future beings. To make it clearer what is at issue here, we can make use of a distinction of A. N. Prior's, between *general* facts and *individual* (or singular) facts. 'That someone has stolen my pencil is a general fact; that John Jones has stolen my pencil, if it is a fact at all, is an individual fact' (Prior 1968: 12). General facts, as we might put it, concern nothing more than the instantiation of certain properties. Individual facts concern *that very person* (or object, or point in space, etc.). Now the unreality of future beings, i.e. those which have not yet come into existence, may consist in this: there are no individual facts about them, only general ones. Interestingly, Prior took this view, not only about future beings, but about *past* beings too. Of Queen Anne, for example, Prior held that

It *was the case that someone* was called 'Anne', reigned over England, etc., even though *there is not now anyone* of whom it was the case that she was called 'Anne', reigned over England, etc. . . . . the fact that Queen Anne has been dead for some years is not, in the strict sense of 'about', a fact about Queen Anne; it is not a fact about anyone or anything—it is a *general* fact. (Prior 1968: 13)

2. Tense before McTaggart: Aristotle, Augustine, and Aquinas

(a) Aristotle

If McTaggart was the first to see clearly the implications of our two ways of ordering events in time, there are at the very least suggestive passages in ancient and medieval writers which can be seen as antecedents, not just of the A-series/B-series distinction, but of the tensed/tenseless debate also.

Aristotle's *Physics* contains a highly sophisticated discussion of two key issues: the continuity of time and the relationship between time and change. His remarks remain highly relevant to contemporary debate in the philosophy of time, and it is not surprising that difficulties concerning the passage of time should have been addressed by him. Indeed, it is possible with hindsight to read into his remarks an anticipation of Russell's theory of time. For when he first introduces the concept of time, in book IV, he begins by drawing out its paradoxical nature, and in order to do so, he makes use of a tensed, A-series formulation: 'Some of it has been and is not, some of it is to be and is not yet. . . . But it would seem to be impossible that what is composed of things that are not should participate in being' ($217^b32–218^a2$). But later, when he presents his own preferred characterization, he uses what seem to be purely tenseless, B-series terms: 'For that is what time is: a number of changes in respect of the before and after' (219^{b}1). Is this an invitation to abandon the tensed view of time, with its paradoxical implications, and embrace a tense-less conception? I think we can be fairly sure that this is not the proposal. First, this would imply a greater division between reality and experience than Aristotle is elsewhere prepared to countenance. Secondly, before presenting his definition, he introduces the 'before' and 'after' in tensed terms:

we say that time has passed when we get a perception of the before and after in change. We mark off change by taking them to be different things, and some other thing between them; for whenever we conceive of the limits as other than the middle, and the soul says that the nows are two, one before and one after, then it is and this it is that we say time is. (*Physics*, IV, 219^{a}26–9)

So the 'before' and 'after' represent aspects of our *past* experience, not change-less relations between events in the world. Thirdly, in his famous discussion of tomorrow's sea battle in *De Interpretatione*, he seems to be recommend-ing a solution to the problem of fatalism which involves an ontological dis-tinction between past and future, a distinction quite at odds with the tenseless, B-series view.

It is worth taking a more extended look at the sea-battle discussion. He is concerned with the question of whether the Principle of Bivalence—that every assertion must be either true or false—can be applied to statements about the future. The apparent consequence of so applying it is that what will happen will happen of necessity:

But if it was always true to say that it was so, or would be so, it could not not be so, or not be going to be so. But if something cannot not happen it is impossible for it not to happen; and if it is impossible for something not to happen it is necessary for it to happen. Everything that will be, therefore, happens necessarily. (*De Inter-pretatione*, 9. 18^a)

This then motivates the fatalist conclusion that, since statements about our future actions will similarly be necessarily true or necessarily false, any impres-sion we have that we have a genuine choice over what to do is an illusion. Since he does not want to accept this conclusion, Aristotle sets about dis-mantling the argument. What he does in fact is to provide *two* solutions: one denies the truth of one of the premises (namely, the Principle of Bivalence); the other shows the argument to rest on a fallacious step in modal reason-ing, from 'Necessarily (p or not p)' to 'Necessarily p or necessarily not p'. So the failure of the argument is, it seems, overdetermined. This has led one commentator at least (Ackrill 1981: 21) to raise a doubt over whether Aristotle is endorsing both, or just one, of the solutions. The conclusion, how-ever, seems to make it pretty clear that he does reject one of the premises of the argument, namely that future-tense statements have a determinate truth-value: 'Clearly, then, it is not necessary that of every affirmation and opposite negation one should be true and the other false' (*De Interpretatione*, 9. 19ᵃ). This implies an ontological asymmetry between past and future: there is no determinate future, as there is a determinate past, to make statements about the future true or false. Since, as we noted above, this asymmetry cannot be captured in B-series terms, Aristotle could be thought of as endorsing a tensed conception of reality, even anticipating Broad's theory. But ascribing such a view to him can only be partially meaningful given that Aristotle did not have a clear conception of the opposing viewpoint. To contemplate future-tensed statements as having a determinate truth-value is not necessarily to see the world in purely B-series terms.

Did Aristotle see a contrast between A-series and B-series? The closest he appears to get to such a distinction is when, at the beginning of the discussion in the *Physics*, he presents us with the following conundrum: 'it is not easy to see whether the now, which appears to be the boundary between past and future, remains always one and the same or is different from time to time' (218ᵃ8). Aristotle's tendency to introduce distinctions might well incline us to expect that he will go on to propose that there is a sense in which the now is always the same and another sense in which it is always different. In McTaggart's terminology, the sense in which the now is always the same is that all uses of 'now' mark the same position in the A-series. It is precisely the boundary between past and future. But the sense in which the now is always different is that it corresponds to different positions in the B-series. That is, different instants, individuated by means of dates, become present successively. This solution exposes the misunderstanding on which the following difficulty rests:

Yet it is not possible either that the same now should persist. For . . . if both previous and subsequent [nows] are in this present now, then events of a thousand years ago will be simultaneous with those of today and none will be either previous or subsequent to any other. (218ᵃ21–9)

We need, in order to see our way out of this difficulty, to recognize the ambiguity in the phrase 'the now'. It can be used in a static sense, to pick out a particular instant which could equally well have been picked out in B-series terms (e.g. '9.00 a.m. GMT on 9 September 1999'). In that sense of 'now', the now does not persist. If it did, then all events would indeed be simultaneous: they would occupy the same instant. But when 'the now' is used in a shifting sense, to pick out an A-series position, the relationship with simultaneity is more complex. If it is now the case that *p* and it is also now the case that *q*, then *p*'s being the case is simultaneous with *q*'s being the case.

However, to *have been* (correctly describable as) 'now' is not to be simultaneous with what is now the case. Non-simultaneous events can occupy the same position in the A-series, provided they do so at different times in the B-series.

Richard Sorabji has argued persuasively that Aristotle did not see the distinction which would have allowed him to dissolve these conundrums, for if he had he would not have offered instead a much more complex, obscure, and unsatisfactory account. Moreover, his use of 'the now' oscillates between both the static and shifting senses, yet he provides no evidence that he is aware of the tension between these senses. (See Sorabji 1983: ch. 4, especially 48–9.)

We cannot find an anticipation of McTaggart here, then. But what we can say is that Aristotle raises just the kind of difficulties which require the distinction between the two kinds of series, and it would not at all have surprised us if he had made that distinction.

(*b*) *Augustine*

St Augustine's *Confessions* contains the most famous discussion of time in medieval literature, if not the most famous discussion of time *tout court*. Here we find a clear antecedent to Broad's and Prior's conception of the unreality of the future (and, I would say, to Prior's conception of the unreality of the past), though the role of the mind in Augustine's account has no counterpart in these later writers.

Time raises tricky theological difficulties for Augustine. In particular, he is exercised by the question of how to reconcile the immutability of God with the Creation as occurring at a particular point in time. If he is immutable, the objection goes, then his will never changes. How, then, do we account for the fact that aeons of time passed before God created the universe? Why, if his will never changes, did he create it then? Augustine's answer to this problem is to put God outside time, so that 'although you are before time, it is not in time that you precede it' (*Confessions*, book XI, section 13). This now leads him into a discussion of the nature of time. What is not entirely clear in the account he gives is whether he takes himself to be denying the reality of time. He does, in fact, hint at this, when he says that neither the past nor the future exists, for one no longer is and the other is not yet, and the present is something whose essence is to be continually going out of existence. It seems that of the three components of time, past, present, and future, none of them has any substantial claim to reality.

However, Augustine is not satisfied with this conclusion, for he wishes to account for our experience of time, in particular our experience of things as having a longer or shorter duration. In sections 26 and 27 of book XI, he sets us the following conundrum (I have, of course, considerably compressed his presentation of it):

We measure time (e.g. the duration of certain sounds).	(1)
We cannot measure what does not exist.	(2)
We cannot measure what has no duration.	(3)
Neither the past nor the future exists.	(4)
The present has no duration (for if it did, part of it would be, not present, but past or future).	(5)

(2) and (4) imply:

We cannot measure what is past or future.	(6)

(3) and (5) imply:

We cannot measure what is present.	(7)

The problem is to reconcile propositions (1) to (5). Together they seem to imply that time does not actually consist of past, present, and future, for we cannot measure these, yet we can measure time. But what else could time consist of? Augustine's solution is to say that 'I must be measuring something which remains fixed in my memory. It is in my own mind, then, that I measure time. I must not allow my mind to insist that time is something objective' (ibid., section 27). How is this a solution? It offers us a reformulation of (4) as follows:

Neither the past nor the future exists independently of our minds.	(4')

Provided that we do not read (2) as insisting that the only things we can measure are those things which exist independently of our minds, then (2) and (4') do not imply (6). Hence we can avoid the unwanted conclusion that time does not consist of past, present, and future.

A simpler solution to Augustine's conundrum is to give up (4) and allow the past objective reality. (We do not need to allow the *future* objective reality, since we clearly cannot measure something which has not yet occurred.) But an earlier remark of his suggests an ingenious argument against this move:

If the future and the past do exist, I want to know where they are. I may not yet be capable of such knowledge, but at least I know that wherever they are, they are not there as future or past, but as present. . . . So wherever they are and whatever they are, it is only by being present that they *are*. (Ibid., section 18)

This is a remarkable passage. Is this the first mention of McTaggart's premiss that every event has all A-series positions? And did Augustine, like McTaggart, see a contradiction in this notion? It is tempting to read the following line of argument into Augustine's brief remarks: if the past (or future) is real, then from the perspective of that past, what to us is past is actually present. So either there is a contradiction here (for how could something be both really past and really present?) or pastness and presentness is simply a matter of perspective. We are probably overreaching ourselves here, but it is interesting to note that Prior, the father of modern tense logic, found in this passage hints of a tense-logical calculus.[1]

(c) Aquinas

If, with hindsight, we can view Aristotle's conception of time as a tensed conception, one which affirms the reality of the A-series, can we find any precursors of the Russellian, tenseless, view? Augustine's remarks, quoted above, are suggestive, but hardly unambiguous. There seem, in fact, to be competing strands in his thought: one is a tensed view in which only the present is real; the other is surely a tenseless view in which the only true reality is eternity, and the flow of time an illusion. It is not clear how he reconciles these. However, medieval discussions of eternity do, I suggest, provide one of the most promising places to look for antecedents to the tenseless view. A long-standing position in Christian thought, defended, as we saw above, by Augustine, but also by Boethius, Anselm, and Aquinas, represents God as immutable. Now if he is immutable, his perceptions (if that is the right word) are similarly immutable. When he contemplates this world, then, does he see it in terms of past, present, and future? Presumably not. But if he sees it as temporally ordered, this order must present itself as akin to a spatial ordering—as Russell's B-series, or McTaggart's C-series, in fact.

The question is whether we can locate this line of thought in any of the theologians who discuss divine immutability and timelessness. It seems to me that Aquinas gets very close to it in the *Summa contra Gentiles*. In describing God's relationship to time, Aquinas makes use of a spatial analogy: we can represent the events of the world as being laid out along the circumference of a circle. God himself is at the centre of the circle. This gives him a perspective on events quite different from that of an observer who is in time.

---

[1] See Prior (1967: 9) and also Findlay (1941), who discusses Augustine at some length. Findlay was the first writer to make explicit reference to the idea of a system of tense-logical laws, and because of this Prior describes him, in *Past, Present and Future*, as 'in a sense the founding father of modern tense logic' (Prior 1967: 1). The first chapter of that book is a useful historical survey of the development of tense logic and its antecedents.

The centre, which is outside the circumference, is directly opposite any designated point on the circumference. In this way whatever is in any part of time coexists with what is eternal as being present to it even though past or future with respect to another part of time. (Aquinas, *Summa contra Gentiles*, book 1, ch. 66)

God, then, does not see things as past, present, or future, but as eternally present, or, as Aquinas puts it in one place, 'God sees all things simultaneously' (ibid., ch. 55). Aquinas's use of the term 'coexists', with its temporal implications, is of course misleading, but the important point is that, if we represent God as immutable, he cannot view time as consisting of a past, present, and future. This raises the intriguing question: if God sees things *as they really are* (and surely, being omniscient, he does), then is Aquinas suggesting that reality does not consist of a past, present, and future? And, if he is suggesting this, does he take it to imply, as McTaggart does, that time is unreal, or is he instead advancing something like Russell's view?

In trying to answer these questions, we cannot do better, I think, than consider Aquinas's solution to an objection which he imagines (or perhaps reports) being raised against the notion of divine eternity. The objection is closely related to the one discussed by Augustine, and goes as follows: if God is immutable, then his will does not change. And since what he wills must necessarily obtain as he wills it, it seems to follow that nothing that occurs as a result of his will can change. For supposing that it did change, and something came to be that was not always the case. Then there would be a gap between God's willing something and that thing coming to be. 'One might conclude further, if God's willing is unchangeable, just as the divine being, and if He wills nothing but what actually is, that he wills nothing but what always is' (ibid., ch. 79). Aquinas's solution to this problem is to say that God does not simply will, and thereby determine, that something should be. He also wills, and so determines, the *time* at which that thing comes to be. There is no gap between God's willing a thing and that thing's coming to be, not because things that are have always been and will always be, but because, for God, all things are present. Now a very persuasive case can be made for the view that the time which God wills each thing to occur must be a B-series time. What, in effect, he wills, is that certain events should occur and that they should stand in certain B-series relations to each other. He does not will that they occur in the past, present, or future, because there is for God no past or future (or present, in the sense of a division between past and future).

However, as we saw above in discussing Aristotle, it is one thing to produce a coherent representation of someone's thought which makes use of a certain distinction. It is quite another to attribute awareness of that distinction to the thinker in question. Rational reconstruction is not (or not necessarily) exegesis. But our brief glance at Aristotle, Augustine, and Aquinas has shown that the intellectual pressure for the distinction made so clearly by McTaggart is indisputably there in writings which antedate his by a matter of centuries.[2]

3. *The new tenseless theory*

(a) *Meaning and truth-conditions*

From the thirteenth century we now move rather abruptly forwards in time to the early 1980s, and the first statements of what has come to be called the new tenseless theory of time. We noted above that Russell sought to eliminate tense from sentences by offering an equivalent tenseless sentence. In 'On the Experience of Time', this took a token-reflexive form. Elsewhere (see Russell 1903), he suggests that a tensed statement is incomplete, and in order to complete it, one needs to know the time at which it was uttered. Thus a fully determinate proposition concerning when something occurred would take the form 'x is F at t', where t is a B-series time. In *The Structure of Appearance*, Goodman proposes that a tensed statement is translated by a tenseless one containing a date.

What characterizes the *old* tenseless theory is the notion of translation: tense can be eliminated from statements because they can all be translated into tenseless statements. What in part characterizes the *new* tenseless theory is the concession that this kind of translation cannot be done. The problem, as John Perry pointed out in 'The Problem of the Essential Indexical' (Perry 1979), is that tense plays a crucial role in thought and action. I will act on the belief that the meeting starts *now*, but not necessarily on any tenseless replacement for that belief, such as that the meeting starts at 12 noon on 5 April 1996. What makes the new theory a tenseless theory, however, is the insistence that the *truth-conditions* of tensed statements are to be given in tenseless terms. So, in terms which reflect McTaggart's language, if not his thought, all in reality that is required to make A-series statements true is B-series relations between events. Of the original formulations of the new theory, one (Mellor 1981) was token-reflexive:

---

[2] The connection between the immutability of God and the unreality of the A-series is made very explicit in Padgett (1992). He also provides an illuminating historical discussion of the doctrine of immutability, but stops just short of attributing a B-series view of time to any of the medieval theologians he considers. The question of whether Aquinas actually held a B-series view is discussed by Craig (1985) (see also Craig's contribution to this volume). The classical discussion of the tension between divine immutability and knowledge of tensed truths (and perspectival truths in general) is Kretzmann (1966). For a discussion of the relation between divine timelessness and the status of the future, see Helm (1988).

The tenseless token-reflexive theory

> Any token *u* of '*e* is occurring now' is true if and only if *u* is simultaneous with *e*.

and the other (Smart 1980) was a date version:

The tenseless date theory

> Any token of '*e* is occurring now', tokened at *t*, is true if and only if *e* occurs at *t*.

A sentence is 'tokened' if it is uttered, inscribed, or otherwise entertained. Note that, according to the date theory, the truth-conditions are independent of the token. This difference between the two tenseless theories is important, as we shall see later. For the moment, we shall be concerned with what the two theories have in common.

At first sight, the new tenseless theory, in either of the versions presented above, seems to drive a wedge between meaning and truth-conditions. But if, as Davidson (1967) and others have argued, meaning is explicable in terms of truth-conditions, does the new theory not simply collapse back into the old? To see how the difference between the two theories can remain consistent with Davidson's project, we need to keep in mind the crucial distinction between type and token sentences. Both the old and new theories can hold that the meaning of a tensed sentence *type* is entirely definable by a tenseless truth-conditional schema, as in the examples above. To put it in other terms, to understand the semantic role played by 'now' in all the sentences in which it occurs, we need only understand the tenseless rule governing its use. What the old theory, as stated by Russell, Goodman, and Gale, holds in addition to this is that the meaning of a tensed sentence *token* is captured by some tenseless sentence. It is this that the new tenseless theory denies. According to the new theory, no tensed token can be equivalent to a tenseless token, because the former, unlike the latter, draws attention to the context in which it is tokened.

(b) Inference

If the new theory *is* properly concerned, then, with the meanings of tensed sentences, understood as types rather than tokens, should it also be concerned with something which seems to depend on those meanings, namely the inferential patterns of tensed sentences? It has been argued (by Oaklander 1990) that a theory about the nature of time itself can be independent of an explanation of the properties of, including the entailment relations between, temporal statements. Against this, Smith (1994) has urged that the tenseless theory of time is explicitly presented by Mellor and others as giving the truth-conditions of temporal (in particular, tensed) statements, and this project must be bound up with an account of how some of those sentences entail others. It is hard to see how Smith's point here can be avoided, especially if we define valid inference in semantic terms as a truth-preserving operation.

To make the issue a little less abstract, I shall consider a difficulty that has been raised for the token-reflexive account. (The problem is essentially the one identified in Smith 1987, but I shall depart somewhat from his presentation of it.) Consider the following two sentences:

It is now 1995.   (1)

1995 is present.   (2)

We intuitively take these to be logically equivalent: each entails the other. So whenever (1) is true, so is (2), and vice versa. (Smith, in fact, takes (1) to imply (2), but not vice versa: see 4(c) below.) Now this implies a relationship between their truth-conditions. Whenever the truth-conditions of (1) obtain, the truth-conditions of (2) also obtain. This is respected by the date version of the tenseless theory, for the truth-conditions of any token of (1), uttered at t, are that t is 1995. Similarly, the truth-conditions of any token of (2), uttered at t, are that t is 1995. As types, (1) and (2) have the same truth-conditional schema, and this means that any simultaneous tokens of (1) and (2) will have the same truth-value.

The inferential connection between (1) and (2) does not, however, appear to be respected by the token-reflexive tenseless theory. Consider a token, u, of 'It is now 1995.' The truth-conditions for u are as follows: u occurs in 1995. Now consider a token, v, of '1995 is present.' The truth-conditions for v are as follows: v occurs in 1995. Consequently, u and v have different truth-conditions: one may obtain without the other. It would seem, then, that the truth-value of any token of (1) is quite independent of the truth-value of any token of (2). If this is so, then the token-reflexive account fails to explain the entailment relations between (1) and (2). The difficulty here, however, is only apparent. Notice first that the truth-conditional schemata for (1) and (2) will have exactly the same form on the token-reflexive account: it is true both of (1) and (2) that any token of it will be true if and only if that token occurs in 1995. This means that, if any given token of (1) is true, then any simultaneous token of (2) will also be true, and vice versa. Of course, the truth of a token of (1) does not entail that there be a true token of (2), but it does entail the truth of any simultaneous token of (2), if there is such a token. That is,

A token, *u*, of 'It is now 1995' is true. (1)

A token, *v*, of '1995 is present' is simultaneous with *u*. (2)

Therefore: *v* is true.

is a valid argument, given the token-reflexive truth-conditions of *u* and *v*. For given these truth-conditions, *u* can only be true if it occurs in 1995. Since it is true, it does occur in 1995, and since *v* is simultaneous with it, *v* also occurs in 1995. Finally, since occurring in 1995 guarantees *v*'s truth, *v* is true.

4. Mixed accounts

(a) Mellor: tensed facts would provide non-token-reflexive truth-conditions

So far, we have tried to characterize, in some detail, the tenseless theory of time, in its various guises. But what of its rival, the tensed theory? Apart from an assertion of the reality of McTaggart's A-series, and an endorsement of his account of change, what does the tensed theory amount to? In particular, what semantics does it provide for tensed sentences? Like the tenseless theory, the tensed theory has undergone a number of changes. And, like the tenseless theory, it now tends to be stated in terms of truth-conditions, although, ironically, the first statement of the tensed theory in these terms was made by a tenseless theorist, Hugh Mellor. According to Mellor (1981), if there is an A-series or, in his terms, there are 'tensed facts', the role of these facts is to provide non-token-reflexive truth-conditions for true tensed statements. His representation of the tensed theory is as follows:

The pure tensed theory

> Any token of '*e* is occurring now' is true if and only if *e* is occurring now.

I call this the 'pure' tensed theory to distinguish it from the more complex accounts which borrow some features from the tenseless accounts presented in the previous section, and which in consequence I call 'mixed' accounts.

Mellor's argument against the (pure) tensed theory goes as follows (see Mellor 1981: 98–101; 1993: 56–8). If the tensed theory is correct, then all tokens of a tensed sentence, such as '*e* is past', are made true by the same fact, namely that *e* is past. So, if that fact obtains, all such tokens are true, irrespective of when they occur. For example, suppose on Monday I utter the sentence 'It is Tuesday.' On the tensed account, my utterance is true only when it is (now) Tuesday. So on Monday, my token is false. Suppose, however, that Monday was yesterday, and it is now Tuesday. Since the tensed truth-condition for my utterance now obtains, that utterance is now *true*. So it is a consequence of the tensed theory that *a token of 'It is Tuesday', uttered on Monday, becomes true on Tuesday*. That is, a tensed token *changes* truth-value over time. We believe, however, that tensed tokens do not change truth-value over time. To pursue the example, we naturally suppose that a token of 'It is Tuesday', if uttered on a Tuesday, is true and remains true. If uttered on any other day of the week, it is false and remains false. The explanation of this is that the truth-conditions of any token of 'It is Tuesday' are that the token occurs on a Tuesday. These are its tenseless token-reflexive truth-conditions (token-reflexive because whether they obtain depends on when the token in question is uttered).

We could represent Mellor's argument, perhaps rather unsympathetically, like this:

> The tensed theory is incompatible with tenseless theory. (1)
>
> The tenseless theory is true. (2)
>
> *Therefore*: The tensed theory is false.

This would, of course, invite a charge of question-begging: we cannot simply assume that the tenseless theory is true in demonstrating the falsity of its rival. Just this charge has been made by Graham Priest (1986), who points out that the incompatibility between the two accounts no more favours the tenseless theory than the tensed theory. What has not been shown by these considerations, it can be urged, is that the tensed theory is self-contradictory. However, premiss (2) does have an independent motivation, that the tenseless theory alone explains our intuitions about the truth-value of tensed tokens.

The tensed theorist has two choices available to him. He can stick to the pure version of his theory and accept the consequence that the truth-value of a tensed token is entirely independent of when that token occurs, perhaps also questioning the idea that this is at all counter-intuitive. Alternatively, he can attempt to preserve Mellor's insight that the truth-value of a tensed token depends on when that token occurs and yet also combine this consistently with the notion that the truth-conditions of that token are tensed. We shall now look at accounts which take this second line.

(b) *Lowe: tensed token-reflexive truth-conditions*

The basic form of the tenseless token-reflexive theory, we recall, is this:

> Any token *u* of '*e* is occurring now' is true if and only if *u* is simultaneous with *e*.

What makes this a tenseless theory is that the two occurrences of 'is' are treated as tenseless, rather than as present-tensed, just as the 'is' in 'The sum of a triangle's internal angles is 180 degrees' is tenseless, and not equivalent to 'is now'. But what if we read the two occurrences of 'is' as present-tensed? We then get a mixed account of the truth-conditions of tensed tokens:

> Any token *u* of '*e* is occurring now' is *now* true if and only if *u* is *now* simultaneous with *e*.

Or rather, we get a mixed account of the conditions for the *present* truth of present-tensed tokens. To complete the picture, we need to add the corresponding schemata for past truth and future truth:

The tensed token-reflexive theory

> Any token *u* of '*e* is occurring now' *was/is/will be* true if and only if *u* *was/is/will be* simultaneous with *e*.

This is the account provided by Jonathan Lowe, in his essay for this volume (though see Section (*d*) below). Although this account is not offered by Lowe explicitly in response to Mellor's argument, it does challenge a supposition of that argument, namely that tensed facts must provide *non*-token-reflexive truth-conditions. Now this mixed account has over the pure tensed theory the advantage that it shows how the truth-value of a tensed token depends in part on its temporal relationship to the event or state of affairs referred to by the token. Does Lowe's account therefore avoid the counter-intuitive consequences of the pure account? Consider again my Monday utterance of 'It is now Tuesday.' Since that token utterance was not, is not, and never will be simultaneous with Tuesday, it was not, is not, and never will be true. It would appear, then, that Lowe's account does avoid the consequence in question, and so escapes Mellor's argument against tensed facts.

There is, however, a question raised by this account: what is the relation between simultaneity (and other 'tenseless' relations) and tensed attributes? Lowe's account rather suggests that tensed attributes are second-order properties, that is properties of the simultaneity relation. This is unlikely to be Lowe's intention, since tensed theorists typically seek to reduce facts about tenseless relations to tensed facts (Tooley 1997 is an exception). The difference between the tenseless and tensed token-reflexive theories is that, where the former makes use of a unitary, tenseless simultaneity relation, the latter uses three tensed relations: past simultaneity, present simultaneity, and future simultaneity. For a different treatment of simultaneity, we may turn to Quentin Smith's suggestion.

(c) *Smith: tensed and tenseless truth*

In *Language and Time*, Quentin Smith introduces a distinction between the present-tensed truth of a token and its tenseless truth. The tenseless truth-conditions of the token determine its tenseless truth-value, and the tensed truth-conditions determine its present-tensed truth-value.

It might appear that this account faces a dilemma: either it introduces a quite unwarranted distinction between two kinds of truth, when truth is essentially a unitary concept, or, if it does not do this, then it faces the kind of contradiction identified by Mellor. That is, there will be times when the tenseless truth-conditions of a token make it true, and the tensed truth-conditions make it false, and either this is a contradiction or truth is an ambiguous concept.

Does Smith face this dilemma? Note first that Mellor represents the putative tensed truth-conditions of a token as something over and above its tenseless ones, but on Smith's account, the tenseless truth-conditions actually follow from the tensed ones:

> Any token *T* of 'The storm is approaching' is *presently* true if and only if
>
> (*a*) *T* is present.
>
> (*b*) The approach of the storm is present.

If (*a*) and (*b*) both obtain, then the tenseless truth-condition

> (*c*) *T* is (tenselessly) simultaneous with the storm's approach

will also obtain (see Smith 1993: 98–100). So the obtaining of the truth-conditions of the present-tensed truth of *T* is sufficient for the obtaining of the truth-conditions of the tenseless truth of *T*. If we add the truth-conditions for past and future truth, we can state the *necessary* and sufficient tensed conditions for the obtaining of the tenseless truth-conditions of *T*. This makes explicit the relation between tenseless relations and tensed facts.

Like Lowe's account, Smith's account shows how the truth-values of tensed tokens depend on when they occur *vis-à-vis* the items they refer to. Since the truth-conditions of my Monday utterance of 'It is now Tuesday' never occur (though each component of the truth-conditions occurs at some time) the utterance is never true. So it seems that Smith, too, avoids Mellor's argument. Finally, Smith denies that he has introduced two kinds of truth. Truth is a unitary concept, he concedes, but there are two different kinds of truth-ascription: the tensed and the tenseless.

There is a remaining difficulty, however. An important feature of Smith's account is his differentiation between 'now' and 'present'. He would reject the suggestion we made earlier, in Section 3(*b*), that 'It is now 1995' and '1995 is present' are logically equivalent. According to Smith, the first entails the second, but not vice versa. The reason is that attributions of now-ness are token-reflexive, but attributions of presentness are not. This follows from his representation of the truth-conditions of 'The storm is approach-ing.' One component of the truth-conditions of any token of this sentence is that the token in question is present. So the truth-conditions of any token of '*e* is present' must simply be that *e* is present: i.e. the truth-conditions of *this* token are *non*-token-reflexive. Now this raises two worries. One is that it seems somewhat arbitrary to distinguish, in terms of truth-conditions, between '*e* is happening' or '*e* is happening now' on the one hand and '*e* is present' on the other. The other is that, even if we concede the difference, there will still, on Smith's account, be certain tensed statements (namely, those of the form '*e* is present') to which the pure tensed account applies. Consequently, these tokens do change their truth-value over time. So if I say on Monday, 'Tuesday is present,' this token must for Smith become true on Tuesday.

(*d*) The tensed date theory

As well as tensed versions of the token-reflexive theory, there are also tensed versions of the date theory introduced in Section 3. The tenseless version, we recall, goes as follows:

The tenseless date theory

　Any token of '*e* is occurring now', tokened at *t*, is true if and only if *e* occurs at *t*.

What makes this a tenseless account is that the 'is' of 'is true', the 'occurs' of '*e* occurs at *t*', and the time, *t*, are all tenseless. But by construing them as tensed, we obtain a tensed version of the date theory:

The tensed date theory

　Any token of '*e* is occurring now', tokened at A-series position *t*, was/is now/will be true if and only if *e* occurred/is occurring/will occur at *t*.

This position is found in Lowe (1987), although the tensing is not made explicit in his formulation. In Chapter 2 below, Lowe suggests that the tensed date theory is logically equivalent to the tensed token-reflexive theory.

　On this account, the token 'It is raining,' uttered two minutes ago, was true if and only if it was raining two minutes ago. If it was not raining then, then the token was false, and does not become true now even if it has just started to rain. What differentiates this theory from the pure tensed theory is that on the former, but not the latter, tokens at different A-series positions have different truth-values. And what differentiates it from Smith's token-reflexive account is that it does not introduce a species of tensed token which does change truth-value over time.

(e) *Tooley: tensed propositions can be tenselessly true*

Diverse though they are, all the accounts above take genuinely tensed proposi-tions (those ascribing an A-series position to an event) to be characterized by their time-variant truth-value. Even the tensed date theory, which takes a tensed *token* to have the same truth-value over time, achieves this result only by taking the token to express different tensed propositions from time to time. These propositions do not have an invariant truth-value. This conception of tensed propositions has recently been challenged by Michael Tooley (1997). He considers sentences of the following form to express tensed propositions whose truth-value is time-*in*variant:

Event *E* lies (tenselessly) in the present at time *t*. (Tooley 1997: 192)

The constancy of truth-value, of course, is guaranteed by the inclusion of a tenseless date in the sentence. The tensed aspect of the proposition is introduced with the words 'in the present'. Tooley anticipates the objection from tenseless theorists that such a sentence is equivalent to the pure B-series statement

Event *E* is simultaneous with time *t*

by pointing out that the first sentence entails that the world is *dynamic*, and the second does not. In a dynamic world, on his conception, the future is unreal. If Tooley is right, the obvious way of McTaggart's paradox, relativizing A-series position ascriptions to B-series times, is actually a way of preserving, not destroying, the A-series.

5. *Future directions*

So far we have been concerned with the past and present of the debate about tense. In this final section I shall offer a personal view of the directions future discussion might take.

It might be thought that the modern debate about tense, cast as it is in terms of truth-conditions, has rather lost sight of Broad's insight (insight, that is, from the tensed perspective) that the fundamental characteristic of time is the unreality of the future, as compared with the reality of past and present.

It is crucially one's view of the status of the future, it could be said, that makes one a tensed or tenseless theorist. However, although the status of the future is perhaps the most important issue in our intuitive conception of time, still the debate about the future must concern itself with truth-conditions. The denial of future fact, if it is to be consistent with the recognition that some future-tensed statements have truth-values, must involve the thesis that future-tensed statements have present-tensed truth-conditions. Any threat to a tensed account of truth-conditions will *ipso facto* be a threat to the un-reality of the future. Conversely, as we saw above in the section on Tooley's account, one could argue that a proper understanding of the semantics of tensed sentences must be based on an ontological theory about the future.

However, there are certain ontological issues which have been somewhat eclipsed by the debate over truth-conditions. One crucial issue, touched on above, concerns the nature of the B-series relations of precedence and simul-taneity. What view should the tensed theorist take about these? There seem to be three options: reductionism (reduce B-series facts to A-series ones), elimin-ativism (deny the existence of B-series relations), and non-reductive realism (allow the independent existence of both A-series and B-series facts). Of these, reductionism is the most attractive. Tenseless theorists have a good reason to deny the existence of A-series if, as many of them believe, such facts involve a contradiction. But no tensed theorist has made out a case for saying that B-series facts involve a contradiction.[3] On the other hand, to allow independent reality to the two series would seem to entail that either A-series and B-series facts overdetermine the truth-values of tensed judgements, or B-series facts have no role to play in determining those truth-values. Neither of these disjuncts seems particularly attractive. So it is in the interests of both sides to examine the prospects for reductionism. Doubts have been raised about the viability of a tensed analysis of B-series relations (see Oaklander 1996, Le Poidevin 1996, and Tooley 1997), but this is still a relatively under-explored area.

A further ontological issue concerns the reality of the past. Although Augustine took the unreality of the past to be as uncontentious as the un-reality of the future, tensed theorists have been reluctant to embrace the idea that past-tensed statements have present-tensed truth-conditions, with its *Nineteen Eighty-Four* revision-of-history connotations. However, it is not clear that the tensed theorist can avoid a symmetrical treatment of past and future.

---

[3] McTaggart, of course, would have said that the B-series did involve a contradiction, because the B-series is derived from the A-series, which itself involves a contradiction. But, in so far as he denied the existence of an A-series, McTaggart was no tensed theorist. The point is that no one (to my knowledge) has wanted to assert the existence of the A-series, deny the possibil-ity of reduction of the B-series to the A-series, *and* hold that the B-series was contradictory.

Consider the following presentation of McTaggart's paradox.[4] Let *a* be some object undergoing change from being *F* to not being *F*, so that it was the case that *Fa* and it is now the case that Not-*Fa*. Now, suppose that the past is real, or, as we might put it, there are past facts. Among the set of past facts is *Fa*; among the set of present facts is Not-*Fa*. So reality contains two facts, *Fa* and Not-*Fa*, which together form a contradiction. There are two solutions to this conundrum. The first is to treat the facts as tenseless facts: *a* is *F* at *t* and Not-*F* at *t'*. The second is to say that reality does not contain *Fa*, but rather the fact that it *was* the case that *Fa*. This, however, is not a past but a *present* fact. Once we add past facts we get a contradiction. The solution, then, seems to be to restrict reality just to present fact. This position has been described as 'presentism', by analogy with 'actualism', which restricts reality to what actually exists. However, since 'presentism' has been used by Quentin Smith (1993) to name a quite different doctrine, namely the idea that presentness inheres in all states of affairs, I prefer to use the term 'temporal solipsism' for the view that only what is present is real. Temporal solipsism has not been without its defenders. Arthur Prior, I believe, was one (see Prior 1970). But it has received less attention than it deserves if, as I personally think, it is the only way for the tensed theorist to escape McTaggart's paradox.

A third ontological issue concerns the tensed understanding of presentness. Is it best seen as a property of events? A number of tensed theorists have denied this (see e.g. Prior 1967: 18, and Levison 1987: 352) while remaining silent about what alternative metaphysical account they would give. Smith, who castigates these theorists for their ontological reticence, is happy to endorse the view that presentness is an intrinsic property of events (Smith 1993: ch. 5). A similar treatment suggests itself for pastness and futurity. But we may wonder whether there are more complex tensed properties than these. How, for example, would we provide a semantics for statements involving iterated tenses, such as 'My chance encounter with Desdemona *was present*'? Clearly, more is being said here than just that my chance encounter is past. The question is whether sentences containing iterated tenses have more complex truth-conditions than those which contain only non-iterated tenses. That is, we may want to say that what makes true 'My chance encounter with Desdemona *was present*' is not simply:

My encounter has the property of pastness   (S1)

but rather:

The presentness of my encounter has the property of pastness.   (S2)

---

[4] What follows is a summary of the argument presented in Le Poidevin (1991: 33).

Both (S1) and (S2) seem to involve the property of pastness, but this similarity may be only superficial. In (S1) the property attaches to an event. In (S2), it is much less clear what the property attaches to. Is pastness here a second-order property? Does the property of pastness attach to the property of presentness? Or to the state of affairs of an event's being present? There is a problem here, for we want if possible a unitary account of tensed properties, according to which they all attach to the same kind of entity, whatever that is. It is not surprising that Prior, whose tense logic depends crucially on the iteration of tenses, rejected the notion of tenses as properties, for syntactically it is simplest to treat iterated tenses as sentential operators rather than as predicates.

There are three responses to the problem. The first is to take Prior's line and reject tensed properties. Advocates of this response, however, need to say far more than they have done about the metaphysics of tenses. The second is to reject iterated tenses as incoherent. This is the line taken by E. J. Lowe (1987), in response to McTaggart's paradox. The third is to develop a coherent theory of tensed properties which can provide a satisfactory semantics for sentences containing iterated tenses. This is attempted by Quentin Smith (1993) (for criticism see Oaklander 1996). A different approach to Smith's is that of Tooley (1997), who treats tensed properties as relational, denying that this amounts to eliminativism. Here again is fertile ground for inquiry.

Iterated tenses also raise important issues for the tenseless theorist. Certainly the simple schema we used above in Section 4 to illustrate the tenseless date theory would not do for such tenses as the pluperfect. For this, we need something like Reichenbach's (1947) account, in which tensed statements are analysed in terms of the time of the event spoken about, the time at which one is speaking, and a third time, the so-called point of reference. The point of reference will sometimes, but not always, coincide with one of the other two. So, for example, in interpreting 'Simone had met René', we need to understand the reference to a time which is both earlier than the time at which the statement is made, and later than the time at which the meeting took place. However, as Hoerl (1996) has pointed out, an account which takes the point of reference to be determined solely in terms of its relation to the time of speech will not do justice to our intuitions concerning the following kinds of case:

'Why didn't John join the football game?'   (1)
'John was tired'

'Why did John leave the party?'   (2)
'John was tired'

A simple account of the semantics of the indicative sentence in the above examples, namely:

A token of 'John was tired', uttered at t, is true iff John is tired at some time prior to t

will deliver the result that the answers in (1) and (2) could be true even if John was not tired at the time of the football game, or at the time of the party. In these cases, it is much more natural to say that the point of reference is determined by the context which *precedes* the statement in question. The construction of a fully worked-out tenseless theory, then, will need to take account of work on tenses in linguistics.

There is a further respect in which the traditional date tenseless theory is at odds with our intuitions. It entails that if two tensed tokens of the same type are tokened at different times, they therefore have different truth-conditions, and so express different propositions. But we want to allow for the possibility of two temporally separated tokens of, say, 'John was tired,' expressing the same proposition. Hoerl's suggestion is that we should regard the proposition expressed by tensed tokens as determined, not by the actual time at which it is tokened, but rather by its intentional context: i.e. the time represented in thought by the producer of the token. (See Hoerl 1996: ch. 5, for a detailed development both of this suggestion, and of the issue discussed in the preceding paragraph.) We may wonder whether the tenseless theorist is entirely free to pursue this intriguing proposal. Recall that the new tenseless theory concedes to tense theory that tense cannot be eliminated from thought, but insists that the truth-conditions for tensed tokens must be tenseless. Now, as long as the tenseless theorist takes the proposition expressed by a tensed token to be determined by its *physical* context (the time at which it occurs), this separation of thought and truth-conditions seems feasible. But once we introduce the idea of an intentional context, the separation between the two looks as though it might be on shaky ground.

In conclusion, an important moral of the section on mixed accounts is that opting for a token-reflexive or date account of the truth-conditions of tensed discourse does not in itself make one a tenseless theorist. There are coherent tensed accounts which similarly provide token-reflexive, or date-indexed, truth-conditions for (at least some) tensed statements. It seems, then, that reflections on the structure of the truth-conditions for tensed sentences, and whether or not tensed tokens change in truth-value over time, will not settle the debate about real tense. A fresh approach is needed. I have outlined above some issues concerning the ontology of time which the debate could usefully focus on. But, equally, it may be that the right way forward is to look at the benefits, or disadvantages, the tensed or tenseless theories confer on discussions of other philosophical issues. Perhaps the question we should ask is not, which theory is coherent? but rather, which theory is more illuminating? It is that thought which is the inspiration for this volume of essays.[5]

REFERENCES

Ackrill, J. L. (1963) (ed.), *Aristotle's Categories and De interpretatione* (Oxford: Clarendon Press).

—— (1981), *Aristotle the Philosopher* (Oxford: Oxford University Press).

Augustine, St (1961), *Confessions*, ed. R. S. Pinecoffin (Harmondsworth: Penguin).

Bourke, V. J. (1975) (ed.), *Saint Thomas Aquinas, Summa contra Gentiles* (Notre Dame, Ind.: University of Notre Dame Press).

Broad, C. D. (1927), *Scientific Thought* (London: Kegan Paul).

Craig, W. L. (1985), 'Was Thomas Aquinas a B-Theorist of Time?', *New Scholasticism*, 59: 475–83.

Davidson, D. (1967), 'Truth and Meaning', *Synthese*, 17: 304–23.

Dummett, M. (1960), 'A Defence of McTaggart's Proof of the Unreality of Time', *Philosophical Review*, 69: 497–504.

Farmer, D. J. (1990), *Being in Time: The Nature of Time in Light of McTaggart's Paradox* (Lanham, Md.: University Press of America).

Findlay, J. N. (1941), 'Time: A Treatment of Some Puzzles', *Australasian Journal of Philosophy*, 19: 216–35.

Gale, R. (1968), *The Language of Time* (New York: Humanities Press).

Goodman, N. (1951), *The Structure of Appearance* (Cambridge, Mass.: Harvard University Press).

Helm, P. (1988), *Eternal God: A Study of God without Time* (Oxford: Clarendon Press).

Hoerl, C. (1996), 'Keeping Track of Time: Time, Thought and Memory' (D.Phil. dissertation, University of Oxford).

Hussey, E. (1983) (ed.), *Aristotle's Physics, Books III and IV* (Oxford: Clarendon Press).

Kretzmann, N. (1966), 'Omniscience and Immutability', *Journal of Philosophy*, 63: 409–21.

Le Poidevin, R. (1991), *Change, Cause and Contradiction* (London: Macmillan).

—— (1996), 'Time, Tense and Topology', *Philosophical Quarterly*, 46: 467–81.

Levison, A. B. (1987), 'Events and Time's Flow', *Mind*, 96: 341–53.

Lowe, E. J. (1987), 'The Indexical Fallacy in McTaggart's Proof of the Unreality of Time', *Mind*, 96: 62–70.

McTaggart, J. M. E. (1908), 'The Unreality of Time', *Mind*, 68: 457–74.

[5] I am very grateful to William Lane Craig, Paul Helm, Nathan Oaklander, Quentin Smith, and a reader for Oxford University Press for their helpful comments on an earlier draft of this essay.

McTaggart, J. M. E. (1927), *The Nature of Existence*, vol. ii (Cambridge: Cambridge University Press).

Mellor, D. H. (1981), *Real Time* (Cambridge: Cambridge University Press).

—— (1993), 'The Unreality of Tense', in R. Le Poidevin and M. MacBeath (eds.), *The Philosophy of Time* (Oxford: Oxford University Press): 47–59.

Oaklander, L. N. (1990), 'The New Tenseless Theory of Time: A Reply to Smith', *Philosophical Studies*, 58: 287–93.

—— (1996), 'McTaggart's Paradox and Smith's Tensed Theory of Time', *Synthese*, 107: 205–21.

—— and Smith, Q. (1994) (eds.), *The New Theory of Time* (New Haven: Yale University Press).

Padgett, A. G. (1992), *God, Eternity and the Nature of Time* (New York: St Martin's Press).

Perry, J. (1979), 'The Problem of the Essential Indexical', *Noûs*, 13: 3–21.

Priest, G. (1986), 'Tense and Truth Conditions', *Analysis*, 46: 162–6.

Prior, A. N. (1967), *Past, Present and Future* (Oxford: Clarendon Press).

—— (1968), *Papers on Time and Tense* (Oxford: Clarendon Press).

—— (1970), 'The Notion of the Present', *Studium Generale*, 23: 245–8.

Reichenbach, Hans (1947), *Elements of Symbolic Logic* (New York: Macmillan).

Russell, B. (1903), *The Principles of Mathematics* (Cambridge: Cambridge University Press).

—— (1915), 'On the Experience of Time', *Monist*, 25: 212–33.

Smart, J. J. C. (1980), 'Time and Becoming', in P. van Inwagen (ed.), *Time and Cause* (Dordrecht: Reidel): 3–15.

Smith, Q. (1987), 'Problems with the New Tenseless Theory of Time', *Philosophical Studies*, 52: 371–92.

—— (1993), *Language and Time* (New York: Oxford University Press).

—— (1994), 'Smart and Mellor's New Tenseless Theory of Time: A Reply to Oaklander', in Oaklander and Smith (1994: 83–6).

Sorabji, R. (1983), *Time, Creation and the Continuum* (London: Duckworth).

Tooley, M. (1997), *Time, Tense, and Causation* (Oxford: Clarendon Press).

2

Tense and Persistence

E. J. LOWE

Amongst the issues that currently divide philosophers of time, two are particularly prominent and, moreover, seem to be interrelated in some way. One is the question of whether a *tensed* or a *tenseless* characterization of time is ontologically more fundamental. The other is the question of whether objects persist through time by *enduring* or by *perduring* (a question which hinges on whether persisting objects can be said to possess *temporal parts*). It is widely assumed—often without much argument—that the tensed view of time goes naturally with an endurance account of persistence while the tenseless view goes naturally with a perdurance account. Although some eminent philosophers of time do not share this assumption,[1] my main aim in this chapter will be to show that, when the terms in which it is framed are properly understood, the assumption is correct. But I shall also be concerned to present and defend a particular version of the tensed view, distinguishing it from certain alternative versions which I find to be problematic.

1. Tensed and tenseless theories of time

I should begin by briefly explaining how I propose to distinguish between the tensed and tenseless views of time. To a first approximation, then, I shall take it that the tensed view regards the notions of *past*, *present*, and *future* as being indispensable ingredients in our understanding of what time is, whereas the tenseless view holds that the metaphysics of time requires only the deployment of the notions of *earlier* and *later*, together with notions definable with their aid—such as that of *simultaneity*, defined, perhaps, as holding between events which are neither earlier nor later than one another (setting aside, here,

I am grateful to David Cockburn, O. R. Jones, Robin Le Poidevin, and a very thoughtful anonymous referee for helpful comments on earlier drafts of this chapter.

[1] See, for instance, Mellor (1981: ch. 8), where he rejects the idea that persisting objects have temporal parts despite his allegiance to a tenseless theory of time. Contrariwise, Storrs McCall (1994: chs. 2 and 7) attempts to combine a tensed view of time with an acceptance of temporal parts. The latter is likewise the position of Michael Tooley (1997).

any difficulties posed by the Special Theory of Relativity). In other words, then, the tensed view takes McTaggart's 'A-series' terminology to be indispensable for the metaphysics—not just the epistemology and semantics— of time, whereas the tenseless view takes 'B-series' terminology alone to be indispensable for the metaphysics of time. Now, having said this, certain possible confusions and false assumptions still need to be dispelled. Some of the most important of these concern the semantics of tensed sentences, to which I shall shortly turn. But first let me say that by a *tensed sentence* I mean a sentence containing a tensed verb or some equivalent linguistic expression. For instance, 'It was raining here yesterday,' 'It is raining here now,' and 'It will rain in Durham on 12 March 2099' are all tensed sentences. (None of these sentences, it will be noticed, *explicitly* employs any of the terms 'past', 'present', and 'future', but each implicitly involves these notions inasmuch as each contains a verb in the past, present, or future tense.) An example of a *tenseless sentence* would be 'Two plus five is seven' or 'Discretion is the better part of valour.'

However, it has to be said at once that there are certain sentences whose status as tensed or untensed is controversial. Thus, an adherent of the tenseless view of time would doubtless say that a sentence such as 'There is rain in Durham on 12 March 2099' is a *tenseless* sentence, akin to 'Two plus five is seven.' By contrast, I would contend that 'There is rain in Durham on 12 March 2099' can *only* be understood as a tensed sentence: either it is equivalent to the (false) sentence 'There is now rain in Durham on 12 March 2099' (false because it is not now 12 March 2099), or it is an ungrammatical way of attempting to say 'There will be rain in Durham on 12 March 2099', or, finally, it is an abbreviated way of saying 'There was, is now, or will be rain in Durham on 12 March 2099' (an interpretation which I shall explain more fully anon). I do not believe that properties can be tenselessly predicated of concrete historical events, but only of abstract entities which do not exist in time (or space) at all. What it *is* to exist in time is to be a potential subject of tensed predications, in my view. That is, for a thing to exist in time is for some property to *be now* exemplified by that thing, or to *have been* exemplified by it, or to *be going to be* exemplified by it. (Of course, I can happily allow that the 'is' in the phrase 'for a thing to exist in time *is* for some property to be [etc.]' is a *tenseless* 'is', because what is being spoken of is something abstract, namely, the property or condition of *being something that exists in time*.) However, this thesis of mine is obviously controversial, so I cannot lay it down as law in advance of adjudicating between the tensed and tenseless views of time.

I turn next, then, to the semantics of tensed sentences. It is sometimes thought that a defender of the tensed view of time must deny that tensed sentences can be supplied with so-called *token-reflexive* truth-conditions, but I do not accept this.[2] Take the present-tensed sentence-type 'It is now raining in Durham.' I am happy to allow that the truth-conditions of (tokens of) such a sentence may be stated as follows:

> A token, *u*, of the sentence-type 'It is now raining in Durham'
> [is] true if and only if *u* [is] uttered at a time *t* such that it [is]
> raining in Durham at *t*   (1)

—or, alternatively but equivalently:

> A token, *u*, of the sentence-type 'It is now raining in Durham'
> [is] true if and only if *u* [is] uttered simultaneously with an
> occurrence of rain in Durham.   (2)

In stating these truth-conditions, I have placed square brackets around various occurrences of the verb 'is', for the following reason. On my view, although a sentence-*type* is an abstract entity which can therefore be the subject of tenseless predications, sentence-*tokens* are concrete entities existing in time, and hence cannot be the subjects of such predications. Accordingly, when I say that a sentence-token, *u*, *is* true if and only if such-and-such *is* the case, either I must mean that it *is now* true or else what I am saying must be a potentially misleading abbreviation for the *disjunctive tensed predication* 'was, is now, or will be true'. In fact, of course, the latter interpretation is the one I should opt for in this instance (in order to state the relevant truth-conditions in their full generality, rather than just for presently existing sentence-tokens). Thus, where I put square brackets around the verb 'is', I mean this to be an abbreviation for 'was, is now, or will be'. The latter construction is admittedly rather clumsy, which is precisely why I am happy to abbreviate it. What I am insisting upon, however, is that *I*, at least, understand the foregoing statement of the truth-conditions of the tensed sentence 'It is now raining in Durham' as being itself *tensed*, because it makes predications of sentence-tokens, which are concrete, time-bound entities. But I appreciate, of course, that an adherent of the tenseless view of time would want to read each instance of 'is' within square brackets above as being a *tenseless* 'is'.

The reason why I lay so much stress on this point is to make it absolutely clear that there is no inconsistency in adhering to a tensed view of time while also endorsing a token-reflexive account of the truth-conditions of tensed sentences. There would only threaten to be an inconsistency here if one were to suppose that tensed sentences can be given *tenseless* token-reflexive

---

[2] The issue is touched on in Mellor (1981: 101).

truth-conditions. Of course, if one *uses* tenses in giving the truth-conditions of tensed sentences, there is a sense in which such an account of their truth-conditions cannot be fully explanatory of their meaning. But this is a familiar feature of truth-conditional semantics quite generally: for instance, quantifier phrases are standardly *used* in giving the truth-conditions of quantified sentences, on both objectual and substitutional accounts of the semantics of quantifiers. (One says, for example, that 'For some *x*, *Fx*' is true if and only if *there is some object* in the domain of quantification which satisfies the predicate '*F*'.) We have to live with the fact that certain notions are so fundamental that no reductive account of them can be supplied: the notions of existence and identity are plausible candidates, but so too, I would urge, are the notions of past, present, and future. I take this irreducibility claim to be part of what the tensed view of time is committed to.[3]

It may be helpful at this point if I also dispose of another red herring, namely, McTaggart's 'paradox'. Famously, McTaggart (1927: 20) condemned A-series terminology as being incoherent, on the grounds that once we use this terminology we are bound to say that *every* event is past, present, and future, despite the fact that 'past', 'present', and 'future' are incompatible predicates. My own response—in which I agree with such philosophers as C. D. Broad (1938: 313–14) and Quentin Smith (1993: 171)—is to deny that we are faced with even the appearance of a contradiction in the first place, because we are *not* bound to say that every event is past, present, and future. In my view, we are at most bound to say something like the following:

> For any event *e*, (i) it either was, is now, or will be true to
> say '*e* has happened', and (ii) it either was, is now, or will be
> true to say '*e* is happening now', and (iii) it either was, is now,
> or will be true to say '*e* will happen'.   (3)

This is, I admit, pretty much of a mouthful—which is why one may be tempted to abbreviate it in misleading ways. What it amounts to is the claim that for any event *e*, *pastness* can be predicated of it in at least one (but not necessarily more than one) of the three tenses, as also can *presentness* and *futurity*. Clearly, though, if—say—presentness is predicated of an event *e* in the past tense, then it may be that presentness cannot without contradiction *also* be predicated of *e* in either the present or the future tense. That is to say, if it *was* true to say '*e* is happening now', then it may be that it *is not now* and *will not be* true to say '*e* is happening now' (though it *is now* and *will be* true to say '*e* has happened'). But (3) by no means implies the denial

---

[3] Tooley (1997) would take issue with me here, but that is because he favours a non-standard conception of the tensed view of time.

of this and consequently cannot be charged with harbouring a contradiction (see further Lowe 1993).

I should, it is true, acknowledge that matters are slightly complicated by the apparent fact that, even if it *was* true to say '*e* is happening now', then for that very reason it *was also* true to say (a little earlier) '*e* will happen'. But the proper way to respond to this point is to distinguish—as Broad (1938: 273) recognized that we should in any case—between different *degrees* of pastness (and of futurity), something we do not have to do in the case of the present tense. Ordinary English accomplishes this by means of adverbial constructions like 'longer ago'. Thus we can say that although it *was once* true to say '*e* is happening now', *longer ago* it was true to say '*e* will happen'. (Later, however, I shall explain why even this suggestion may need to be modified.) Another minor complication is that, in the case of an event *e* of some duration, presentness *may* in fact be predicated non-contradictorily of *e* in, say, both the present tense and the past tense, though only in virtue of *e*'s having as parts two sub-events, *e'* and *e''*, such that presentness can only be predicated of *e'* in the present tense and can only be predicated of *e''* in the past tense.

2. *Some differences between tensed theories*

It will be noticed that although I spoke just now of presentness being predicated of an event *e* in the past tense, the way I construe this is in terms of saying that it *was true to say* '*e* is happening now'. There is one very good reason why I use this sort of metalinguistic construction, and this is that I do not want to suggest that 'presentness' is some sort of *property* of events. Some advocates of the tensed view of time would indeed want to suggest precisely this (see e.g. Smith 1993: ch. 5), but I cannot agree with them. The problem is that if presentness (and likewise pastness and futurity) is conceived of as being a property of events, then it is difficult to see how it can *either* be a property which an event has always had and always will have, *or* be a property which an event now has, has had, or will have only temporarily—and no other option seems available. The first option appears to be quite untenable: it is surely absurd to say that the Battle of Hastings, say, always was and always will be present. According to the second option, however, presentness is a property which the Battle of Hastings *once* had, but which it has no longer. But how has it come to lose this property? We should not be tempted to liken the case to that of a blue object losing its colour with the passage of time: objects can lose or gain properties because they *persist* through time while undergoing qualitative change—but events do not do this. The

Battle of Hastings *no longer exists* now that it has 'lost' its 'presentness', unlike the blue object which has faded. It is true that persisting objects undergo not only qualitative change, but also *substantial* change, when they cease to exist by losing an *essential* property (as, for example, when a coin is destroyed by being squashed out of shape). But loss of 'presentness' by an event cannot be likened to substantial change, since the latter is the ceasing to exist of something that has persisted for a certain period—and events, once more, do not do this. It will not help, apparently, to think of an event's supposed loss of presentness as being a 'mere Cambridge change'—the sort of change which is undergone by a dead person who ceases to be famous—for this would imply that presentness is not an intrinsic property of events and hence its loss not a real change in the event possessing it. Nor, I think, should we say with Broad that we are talking here of a *special* sense in which properties can be gained or lost by events, a sense in which such 'changes' precisely *constitute* the 'passage of time'—for this is to indulge in a kind of talk about time which, though common enough, strikes me as being hopelessly obscure.[4]

However, now the charge may be raised against me by other adherents of the tensed view of time that in denying that presentness (and likewise pastness and futurity) is a *property*, I deny myself any means of explaining the distinctive semantic contribution of tenses to our talk about time (see e.g. Smith 1993: 166 ff.). What does an expression like 'the present moment' *mean*, if it does not involve ascribing the property of presentness to a moment of time? My response is twofold. First, not every adjective in the surface structure of everyday grammar should be construed as having the logical role of expressing a *property* of items of some sort. Sometimes such an adjective may effectively be playing the logical role of a sentential operator or predicate modifier. For example, possible-worlds semanticists typically claim that 'Possibly *P*' means 'In some possible world, *P* is true'—but one might be (in fact, I am) inclined to reverse the direction of explanation here, and say that this use of the adjective 'possible', if it is to be understood at all, is to be construed *not* as expressing some property of items of a queer sort ('worlds'), but rather in terms of the sentential operator 'Possibly'. Likewise, then, 'There is rain in Durham at the present moment' is, in my view, just a roundabout way of saying 'Presently, it is raining in Durham'—or, even more succinctly, 'It is now raining in Durham.' My second point is to repeat what I said before about certain notions being so fundamental that they are

---

[4] C. D. Broad and Storrs McCall both maintain that we can legitimately use the language of change to characterize temporality provided we distinguish between change *in* time and change *of* time: see Broad (1923: 64 ff.) and McCall (1994: 30–1). This suggestion is, I think, unhelpful, because there can be literally nothing in common between the two sorts of 'change': so why use the same word for both?

semantically irreducible—I take this view of the tenses. And why not? Do we not have to take the same view of an operator such as *negation*? 'It is not the case that' cannot be given a non-circular semantic explanation, yet of course we grasp its meaning as clearly as we grasp the meaning of anything.

At this point it may seem that my position with regard to the tenses is pretty much the same as A. N. Prior's, and yet in fact that is not so. This is connected with the fact that I use the metalinguistic constructions deployed earlier. Prior believed that the so-called tense operators could be *iterated*, allowing us to say things such as 'It was the case that it will be the case that such-and-such' (see Prior 1968: 8). But because I believe that tensed sentences have token-reflexive truth-conditions, I cannot accept this sort of locution as legitimate—any more than I can accept as legitimate a sentence like 'Over there it is raining here' (cf. Lowe 1987*a*). The best we can do to make sense of the latter is to construe it as meaning 'Over there it is true to say "It is raining here."' Likewise, then, instead of saying 'It was the case that it will be the case that it is now raining', I must insist on saying 'It was true to say "It will be true to say 'It is now raining.'"' In sum, then, I adopt the metalinguistic constructions exemplified above in order to avoid the difficulties of two versions of the tensed theory of time which I wish to reject: the view which treats presentness, pastness, and futurity as properties of events or moments, and the view which treats the so-called tense operators as iterable.

I should perhaps remark that in fact I differ from Prior also in *not* regarding the tenses as sentential operators, but rather as *predicate modifiers*: they cannot, in my view, be sentential operators because any sentence on which they operated would have to have a *tensed verb*, so that at least *some* tensed sentences could not result from the application of a tense operator to a sentence—and if not these, then why any? Prior himself thought that the basic sentences from which all other tensed sentences were generated by the application of tense operators were *present-tense* sentences (see Prior 1968: 8) —which is, in my view, to show a quite unjustified favouritism towards the present. No wonder, then, that Prior is often accused of 'presentism', an implicit denial of the 'reality' of past and future.[5] No such charge can be levelled at me, precisely because I treat all three tenses as being on an equal footing.

This is not, however, to say that I think (as tenseless theorists of time typically do) that no ontological distinction whatever can be drawn between past, present, and future—an idea that has often been associated (rightly or wrongly) with some sort of fatalism or determinism. I am happy to allow, thus, that the law of excluded middle may not apply to all future-tensed

---

[5] See e.g. Le Poidevin (1991: 36 ff.), where this doctrine is called 'temporal solipsism'. Quentin Smith (1993) uses the term 'presentism' to describe his own position, though he distances his position quite considerably from Prior's.

statements—that maybe, for example, it *is now* neither true nor false to say 'It will rain in Durham on 12 March 2099' or 'There will be a sea battle tomorrow.' (Incidentally, if this is correct, then I should at least qualify my earlier remark that if it *was* once true to say '*e* is happening now', then it *was also* true to say, earlier, '*e* will happen': for in many cases it could be that it was earlier *neither true nor false* to say '*e* will happen', even though it was later true to say '*e* is happening now'.) On this matter of 'future contingents', then, I side with what I take to be Aristotle's position (in *De Interpretatione*, ch. 9). But be that as it may, returning to the question of the status of the tenses in logical grammar, I repeat my view that they are *predicate modifiers*. To say that Durham will be rainy is to predicate raininess of Durham in the future-tense mode. And it is because Durham is a concrete, time-bound object that, in my view, properties like raininess can only be predicated of it in some tensed mode, either past, present, or future.

3. *The tenseless view of time and an objection to it*

I hope I have said enough now to explain my version of the tensed view of time and show that it is at least consistent and intelligible. But what of the tenseless view? What is it and why do I reject it? I do not have space to go into my reasons for rejecting it in much detail here, since my primary concern is to defend the tensed view and show that it commits me to an endurance account of persistence (of which more in due course). As for what the tenseless view *is*, I start from my characterization of it in Section 1, as the view that the metaphysics of time need have recourse only to B-series terminology, that is, to an account of time in terms of earlier/later relations between events or moments, together with relations like simultaneity which are apparently definable in terms of earlier/later relations. I speak disjunctively here of events *or* moments, because some tenseless theories of time are 'substantival' theories, which regard moments of time as fundamental entities in their own right, while other tenseless theories of time are 'relational' or 'modal' theories, which attempt to construct moments of time (if they acknowledge their existence at all) out of events in some fashion. This is not a distinction with which I am particularly concerned here. I wish to reject all tenseless theories of time, both substantival and relational. Clearly, an essential feature of all tenseless theories is that they must regard tenseless predication as fundamental and hold that all tensed sentences can be given tenseless truth-conditions. (Such theories would, typically, endorse the token-reflexive account of the truth-conditions of tensed sentences exemplified in (1) and (2) of Section 1, but insist that the 'is' in square brackets be interpreted *tenselessly* in each of its occurrences.)

Now, one objection to the tenseless theory of time that I have is that it does not tell us what it is for something to exist *in* time. Recall that, by my own account of Section 1, what it is for something to exist in time is for that thing to be a subject of tensed predications—for it to *have had*, or to *have now*, or to *be going to have* some property.[6] Clearly, an adherent of the tenseless theory of time cannot appeal to any such consideration as this. But what else can he say? Perhaps he could say that for something to exist in time is for it to have (tenselessly) some property *at some moment of time*, or for it to have (tenselessly) some property *simultaneously with some event* (though the latter version will require an account of eventhood which, problematically, does not already presuppose a notion of existence in time). But then it will not be clear why we should not say, for example, that *the number 4* exists in time, because it will not be clear why we should not say that 4 has (tenselessly) the property of being the square of 2 at *every* moment of time, or simultaneously with *every* event. (My own view, of course, is that 4 does indeed have this property tenselessly, but that for this very reason it *does not* have the property *at* any moment of time; however, this is a proposal which is evidently not available for the tenseless theorist to adopt, because he obviously cannot allow that to predicate a property of an object tenselessly is *ipso facto* to imply that the object does *not* possess the property *at* any time, since *all* of his predications are ultimately tenseless.)

Now, if the tenseless theorist responds by saying that the number 4's having (tenselessly) the property of being the square of 2 is not an *event*, or *temporal state of affairs*, and for that reason does not occur 'at' any time and is not simultaneous with any event, then we need to be told what *does* count as an 'event'. It hardly helps to say that a state of affairs is 'temporal', and so 'in time' and an 'event', just in case it *enters (tenselessly) into temporal relations* with other states of affairs (for instance, by being earlier or later than other states of affairs). For, once again, what is then to prevent us from saying that 4's being the square of 2 is a 'temporal' state of affairs in virtue of standing in the temporal relation of simultaneity to *every* other state of affairs? That this would apparently require us to abandon the assumption that simultaneity is an equivalence relation does not seem to me decisive; after all, there is a long tradition of theological thinking, beginning with Boethius, which holds, without obvious incoherence, that God's thoughts are simultaneous with everything that happens at every moment of time. In any case, even if this objection *were* deemed decisive, what would prevent us from saying instead that 4's being the square of 2 is a 'temporal' state of affairs

[6] In order to make this proposal work effectively, we may need to restrict the properties in question to *intrinsic* ones, so as to avoid having to say, for example, that the number 4 exists in time because it *has had* the property of being someone's favourite number.

in virtue of having *successive temporal parts*, each of them earlier than, later than, or simultaneous with some other state of affairs? In short, the proposal now under consideration simply fails to perform the task demanded of it, namely, that of explaining why a state of affairs such as 4's being the square of 2 should not be regarded as an *event* (perhaps an infinitely protracted and boring one) and so as something existing 'in time'.

My own view (again a broadly 'Aristotelian' one) is that events are changes (and unchanges) in the properties and relations of persisting objects, so that to say that some event occurred is to say that some object or objects changed or remained the same in certain ways—the objects in question being, of course, ones *existing in time*. So my account of events already presupposes the notion of existence in time, and thus is not available to the tenseless theorist of time for present purposes. The tenseless theorist may perhaps appeal, instead, to *causation* as a criterion of eventhood, saying that events are states of affairs which have (tenselessly) causal relations to other states of affairs—which might appear to exclude from the class of events the state of affairs of 4's having (tenselessly) the property of being the square of 2. But this presupposes, implausibly, that causation can be analysed independently of the notion of existence in time. On any plausible account, to identify *a* as a cause of *b* is *first* to say of *a* and *b* that they *occur(red)* at certain times, and *then* to say something further about their relationship to one another.[7]

Now, I do not want to claim that the objection which I have just raised to the tenseless theory of time is absolutely insurmountable—obviously, I cannot anticipate every possible rejoinder that might be made to it—but it does seem to me unsurprising that the theory should be open to such an objection. After all, the theory is committed to saying that there is nothing fundamentally different about predicating a property of some abstract, timeless object and predicating a property of something existing in time—in both cases, tenseless predication supposedly suffices. It is unsurprising, then, that the theory should be open to the charge that it offers not merely a *tenseless*, but a *timeless*, view of time, and thereby eliminates the very phenomenon it is supposed to explain.

4. Theories of persistence and the notion of 'temporal parts'

I want to move on now to the problem of *persistence*. As I remarked in my opening paragraph, theories of the persistence of objects through time are

---

[7] Certainly, that is the way in which both David Hume and David Lewis approach the analysis of causation: see Lewis (1986*a*).

commonly divided into two categories: *perdurance* theories and *endurance* theories. Standardly, the theories are characterized as follows.[8] According to the perdurance account, an object persists through time by having different temporal parts at different times at which it exists, whereas according to the endurance account, an object persists through time by being 'wholly present' at each time at which it exists. (I have to say that I find the expression 'wholly present' less than fully perspicuous, but I take it to mean, at the very least, that persisting objects do *not* have temporal parts.) The debate turns, then, on the question of whether or not persisting objects have temporal parts.

Unfortunately, the very term 'temporal part' has a number of different possible meanings, and is very difficult to clarify in any useful sense. One thing we need to be clear about is that the notion of a temporal part of a persisting object is *not*, for the purposes of the perdurance versus endurance debate, to be assimilated to the notion of a *temporary* part of a persisting object. The leg of a chair might be a temporary part of it, if it is a replacement for the original leg while the latter is undergoing repair. But such a chair leg is itself a persisting object, which will itself have temporal parts if all persisting objects do. Of course, some philosophers apparently do hold that at least some temporal parts of persisting objects are themselves persisting objects—as when David Lewis (1983: 76) likens a temporal part of a person to a short-lived person. But, clearly, this is not a view which a perdurance theorist would do well to take about *all* temporal parts, since that would deprive him of any non-circular account of persistence. Sometimes, temporal parts of persisting objects are purportedly introduced to us as entities representable by ordered pairs of objects and times—for instance, by the pair ⟨Napoleon, 5 June 1805⟩—and such an entity, we are led to suppose, might be referred to by a complex noun phrase such as 'Napoleon on 5 June 1805'. Whether one can concoct entities in this easy fashion is a matter for debate, but even if one can, it seems plain that a temporal part thus conceived is both conceptually and ontologically posterior to the persisting object (here, Napoleon) of which it purports to be a part. So jejune a notion of temporal part accordingly seems of little use in a metaphysical account of the persistence of objects—as is demonstrated by the fact that even an endurance theorist need apparently have no special qualms about countenancing the existence of temporal parts thus conceived.

Is there, then, *any* acceptable notion of temporal part that is robust enough to serve a useful metaphysical purpose? Well, it is sometimes urged that *events and processes*, at least, clearly have temporal parts in a robust sense of this

---

[8] See further David Lewis (1986*b*: 202). See also Lowe (1987*b*), in which I criticize an important argument offered by Lewis in support of the perdurance theory. I argue more positively on behalf of the endurance approach in Lowe (1988) and in Lowe (1994).

sort. (I take a process to be a sequence of temporally successive events, the events in the sequence being parts of the process.) We talk, after all, of such items as an *early* part of the performance of a certain play, such as the performance of its first scene. In like manner, we may talk of an early or late part of a persisting object's *existence* or *life*—though here I would agree with those who insist that this does not of itself entitle us to think that there are such entities as early or late parts *of that object*. But still, it may be urged, even if it is indeed open to debate whether such entities exist, at least we have now hit upon an intelligible way of talking about the supposed temporal parts of persisting objects which can be put to use in the metaphysical debate as to whether or not such objects persist by perduring. (If we could not even talk intelligibly about the possibility of objects having such parts, the debate could never get started; though I suppose some might say that that would be no bad thing.)

Now, I do not think it would necessarily be fair to protest that to model the supposed temporal parts of persisting objects upon the supposed temporal parts of processes in this way is simply to *reduce* persisting objects to processes. (Certainly, though, I would want to reject any such reduction because, as I explained earlier, I take an 'Aristotelian' view of events and processes as consisting in changes and unchanges in the properties and relations of persisting objects.) Even so, I have serious doubts about the attempted analogy with events and processes. For I am not sure that even *they* have 'temporal parts' in any metaphysically significant sense. The performance of the first scene of a play is, no doubt, a *part* of the performance as a whole. (Processes, as I said a moment ago, are sequences of events, which are parts of them.) But in what sense is the performance of the first scene a *temporal* part of the whole performance? In the sense that it occurs, or comes into existence, before the rest of the performance? But if *that* were all there were to its being a temporal part, then, it seems, the proposed analogy between the 'temporal parts' of events and the 'temporal parts' of persisting objects would require us to regard the *temporary* parts of persisting objects as 'temporal parts' of them. After all, the replacement leg of a chair, which is a temporary part of it, may come into existence before or after certain other of its parts—so that if all that is required for something to be a temporal part of a persisting object is that it be a part of it and come into existence before or after other parts of it, then the replacement leg is as good an example of a 'temporal part' of the chair as anything could be. Nor will it help much to point out that an 'earlier' part of the performance of a play *ceases to exist* before its 'later' parts begin to exist, whereas the temporary leg of a chair may go on existing even after it has been replaced as a part of the chair—for, clearly, the temporary leg *could* perfectly easily be destroyed in the course of removing it from the chair. Indeed, some temporary parts of persisting objects *have* to be destroyed in the course of removing them from the objects in question: for example, certain living parts of organisms. It is true that, at the time at which the 'later' parts of the performance of a play exist, *none* of the 'earlier' parts still exists. But equally, in the case of a chair, there may be a time at which none of the parts—legs, arms, seat, and so forth—which it had at some former time still exists, all having been destroyed and replaced by newly fashioned parts. Are we then to say that such parts of a chair are *temporal* parts of it? If we do, we shall clearly not be deploying a notion of 'temporal part' which can have any relevance to the debate between perdurance and endurance theorists of persistence, because such parts of a chair are themselves just other persisting objects about the nature of whose persistence the debate is concerned.

My own view is that if any metaphysically significant notion of the temporal parts of persisting objects is to be constructed, then the proper analogy to make is not with items like the performance of the first scene of a play, but rather with the *spatial* parts of concrete things (whether these be persisting objects or events). And here I would point out that by a spatial part of, say, a chair, I *do not* mean something like one of its legs. Rather, I mean a spatially extended entity whose spatial boundaries are defined in relation to the chair: something such as the bottom three inches of one of its legs, or the left-hand half of its back as seen from the front. Clearly, concrete things can only have spatial parts thus conceived if the things in question are *extended in space* (this applies just as much to events as to persisting objects). By the same token, then, concrete things can only have *temporal parts*, in the sense in which I want to speak of them, if those things are *extended in time*.[9] By this account, a temporal part of a persisting object would be a temporally extended entity whose temporal boundaries are defined in relation to that object. (One minor refinement needs to be made at this stage: just as we should, in principle, admit two-, one-, and zero-dimensional spatial parts of things in addition to three-dimensional ones, so too we should admit zero-dimensional temporal parts of things if we admit one-dimensional ones—but nothing crucial hinges upon this point, since what is important is that we must indeed be prepared to speak of temporally *extended* entities if we are to talk of temporal parts in any metaphysically significant sense.)

So, then, a temporal part of Napoleon, if such an entity exists, would be a temporally extended entity whose temporal boundaries are defined in relation to Napoleon—thus the boundaries of one such part might be fixed by

---

[9] Here, I believe, I may be in agreement with Robin Le Poidevin, even though he wants to say that things *are* extended in time whereas I want to deny this: see Le Poidevin (1991: 60 ff.).

the date of his birth and the date of his tenth birthday. One might try to refer to such an entity by some such complex noun phrase as 'Napoleon up to the age of 10'. (This notion of temporal part is, I believe, quite distinct from the 'ordered pair' conception discussed earlier, because the latter conception involved no commitment to the existence of *temporally extended* entities of any sort. And this is not just because each such pair contained only a *single* time: for, in the first place, it was not implied that the time in question had to be *momentary*—the example I gave actually specified a *day*—and, secondly, no material difference would have been made if instead of ordered *pairs* we had talked of ordered *triples*, each containing *two* times, for there would still have been no commitment to the existence of temporally extended entities.) What emerges from all this, if I am correct, is that the perdurance versus endurance debate does not really hinge upon issues in *mereology* (the study of part–whole relations) as such, but rather upon the question of whether *anything*—and here I include not only persisting objects but also events and processes—is *extended in time*, in anything like the way in which things are *extended in space*. But this is at bottom a question about the nature of time, rather than a question about the nature of things existing in it. The question is whether we can properly talk about time as being some sort of *dimension* of reality, relevantly akin to the three dimensions of space. (A word of caution here: there is an abstract or purely formal notion of dimensionality used, for instance, to describe perceptible colours as differing from one another along the three independent 'dimensions' of hue, brightness, and saturation— but in what follows I am concerned with a much more robust notion of dimensionhood for which spatial length, breadth, and height constitute the paradigm, all of these providing independent ways in which concrete things can be *extended*.)

5. *Can anything be extended in time?*

Our question, then, is this: can we properly talk about time as being a dimension of reality in which concrete things are extended? I think that our answer to this should depend on whether we believe that the tensed or the tenseless view of time is correct. If the tensed view is correct, then I think we cannot properly regard time as being such a dimension of reality, and accordingly cannot accept a perdurance account of the persistence of objects, given my understanding of what such an account implies (namely, the existence of temporally extended entities). Now, it is not my current task to demonstrate beyond all question that the tensed view of time is correct. The most I hope to do is to show why the tensed view *cannot*, while the tenseless view *must*, regard persisting objects as being temporally extended.

Since space provides the paradigm of extension, we can conceive of time as a dimension in which things can be extended only in so far as we can find suitable similarities between space and time. The only case that can be made for there being a relevant similarity, it appears to me, rests upon a comparison between the relations between different points on a line and the relations between different times. Now, the tenseless theory of time, with its exclusive reliance on B-series terminology, does indeed make space and time appear relevantly similar. For, on this view, the earlier/later relations between times (and events) suffice to define a 'betweenness' relation on them which is entirely analogous to the betweenness relation of points on a line. (A time *t* is *between* two other times *t′* and *t″* just in case *t* is later that *t′* and earlier than *t″* or *t* is earlier than *t′* and later than *t″*—and similarly with events; moreover, *all* times are thus related to one another, in a unique linear sequence.) Quite crucial here is the fact that the tenseless theory makes *no ontological distinction* between different times, nor between the existence of things *at* different times—for this precisely parallels the spatial case. Of course, there is still the *anisotropy* of time—its directedness—to consider, but this does not appear to be a particularly important difference between time and space as far as the tenseless theory is concerned, or, at least, not important enough to undermine the theory's commitment to an extensional view of time.

There seems, then, to be no good reason why a tenseless theorist of time should regard space, but *not* time, as being a dimension of reality in which things are extended. But matters are, I believe, quite otherwise for the tensed theorist of time. For him the fundamental notions are of past, present, and future, which have no spatial analogues. It is true, no doubt, that the *tenseless* theorist will urge that there is a close parallel between 'now' and 'here': but the *tensed* theorist will vigorously deny this, because while he may happily allow that the truth-conditions of sentences containing spatial indexicals such as 'here' *can* be stated in a metalanguage which does not itself deploy spatial indexicals, he will—as we have seen—hold in contrast that tensed sentences can only be provided with *tensed* truth-conditions. In short, the tensed theorist may agree with the tenseless theorist about the semantics of 'here', but not about the semantics of 'now'.

On the tenseless view, any event *e is tenselessly* between at least two other events *e′* and *e″*—with the possible exceptions of a first and a last event at the ends of time—just as, on this view, any point is tenselessly between at least two others on a line. But on the tensed view, although *e* can be said, for example, to *have occurred less long ago than e′ and longer ago than e″*, this does not license us to say that *e* has or had some 'betweenness' relation to *e′* and *e″* in anything like the sense which obtains in the spatial case. For, on the tensed view, *e′* had already *ceased to exist* when *e* came into existence, and the latter had ceased to exist when *e″* came into existence. Accordingly, we do not have here a relationship between entities which are in any sense *coexistent*, and for that reason not one at all like that between points on a line. Now, it is true that a tenseless theorist will also say that, in one sense, *e*, *e′*, and *e″* do *not* coexist—namely, in the sense that they do not all exist *at the same time*. Yet, for the tenseless theorist these events do still coexist in another sense, inasmuch as it is tenselessly true of all of them that they do indeed *exist*—for, on the tenseless view, '*e* exists at *t*' clearly entails '*e* exists', *simpliciter*. (If you like, for the tenseless theorist all these events coexist, tenselessly, *in the same possible world*—the actual world—and differ not at all from one another in respect of their *ontological status* within that world, but only in respect of their *temporal location* within it.) Clearly, it would indeed be simply mistaken to describe the tenseless theorist as treating all events as being *cotemporal*. But the 'co' in 'coexistent' need not be taken to mean 'at the same time'—it may be taken to mean 'together' in some other sense. And then the point is that for the tenseless theorist another such sense is undoubtedly forthcoming, whereas for the tensed theorist there is simply *no* sense in which *e*, *e′*, and *e″* can be said to coexist, because on this view existence can *only* be ascribed to them in a tensed way—and in that way they cannot be said to coexist.

My claim, then, is that any 'betweenness' relation relevantly akin to that relating points on a line is a relation between items which must in *some* sense coexist—but that the tensed theorist allows no such sense for events separated in time. And this is why I conclude that the tenseless theory alone treats time as a dimension of reality in which things are extended, and accordingly treats persisting objects as having temporal parts (in the only metaphysically significant sense available) and so as *perduring*—that is, as persisting over time by virtue of having different temporal parts at different times. On my version of the tensed view, by contrast, to say that a presently existing object has persisted is just to say that that very object *did* exist in the past and still does so now. There is, on this view, no implication whatever that in order for this very object to have persisted until now, some *other* object (a temporal part of it) must have existed in the past. Notice that, in the spatial case, it *does* seem intuitively right to say that a spatially extended object, such as the earth, exists now *elsewhere* than here (where 'here' refers, say, to where Durham is) by virtue of having spatial parts which now exist in other places (for instance, where San Francisco is); and the same applies, on a smaller scale, to objects like chairs. But, as I say, the verdict of the tensed view is that nothing analogous applies in the temporal case. For what it is worth, I consider it to be a distinct merit of the tensed view of time that it delivers this verdict, for it surely coincides with the verdict of common sense.

REFERENCES

Broad, C. D. (1923), *Scientific Thought* (London: Routledge & Kegan Paul).

—— (1938), *Examination of McTaggart's Philosophy*, vol. ii, part i (Cambridge: Cambridge University Press).

Le Poidevin, R. (1991), *Change, Cause and Contradiction* (London: Macmillan).

Lewis, D. (1983), 'Survival and Identity', in his *Philosophical Papers*, vol. i (New York: Oxford University Press).

—— (1986a), 'Causation', in his *Philosophical Papers*, vol. ii (New York: Oxford University Press).

—— (1986b), *On the Plurality of Worlds* (Oxford: Basil Blackwell).

Lowe, E. J. (1987a), 'The Indexical Fallacy in McTaggart's Proof of the Unreality of Time', *Mind*, 96: 62–70.

—— (1987b), 'Lewis on Perdurance versus Endurance', *Analysis*, 47: 152–4 (repr. in H. Noonan (ed.), *Identity* (Aldershot: Dartmouth, 1993)).

—— (1988), 'Substance, Identity and Time', *Proceedings of the Aristotelian Society*, Supp. Vol. 62: 61–78.

—— (1993), 'McTaggart's Paradox Revisited', *Mind*, 101: 323–6.

—— (1994), 'Primitive Substances', *Philosophy and Phenomenological Research*, 54: 531–52.

McCall, S. (1994), *A Model of the Universe* (Oxford: Clarendon Press).

McTaggart, J. M. E. (1927), *The Nature of Existence*, vol. ii (Cambridge: Cambridge University Press).

Mellor, D. H. (1981), *Real Time* (Cambridge: Cambridge University Press).

Prior, A. N. (1968), *Papers on Time and Tense* (Oxford: Clarendon Press).

Smith, Q. (1993), *Language and Time* (New York: Oxford University Press).

Tooley, M. (1997), *Time, Tense, and Causation* (Oxford: Clarendon Press).

3

Seeing the Present

JEREMY BUTTERFIELD

1. Introduction

Philosophers of time fall into two broad camps. For *detensers*, the present is an epistemological/subjective notion reflecting our limited knowledge at any time of a temporally extended reality, just as where we are limits our knowledge of distant objects (Taylor 1955; Quine 1960: section 36; Mellor 1981). *Tensers* distrust this analogy between the present and the local; for them the present but not the local is an ontological/objective notion (Prior 1959, 1970; Dummett 1960; Gale 1968: 213). This crude contrast is made more precise in various ways. For example, detensers accept the idea of an actual past and future—what in fact has happened and will happen; tensers usually reject the idea of an actual future, and sometimes (Dummett 1968; Prior 1970) of an actual past.

I am a detenser. But there is no doubt that strong intuitions seem to support the tenser's view that the present is objective. Here are three. We more readily take as real the presently existing objects, wherever they are, than the objects that are at some time located here (Putnam 1967; Lloyd 1978: 217). We are more apt to give sentences time-variable truth-values than space-variable ones (Dummett 1973: 386, 390). And we think of ourselves as sharing a common, albeit ever-changing, *now*, while we each have a different *here* (Gale 1964: 105). If all and only presently existing objects (but no objects that are sometime here) are in some sense real, the present is surely objective in a way the local is not. Similarly the second and third intuitions suggest that the present and its 'movement' are in some way objective.

Detensers must explain away such intuitions, i.e. explain why we have them without endorsing them, or without conceding that they support the tenser. This chapter attempts to do that for just these three intuitions. I restrict

I would like to thank Hugh Mellor for comments on earlier versions; and Jonathan Grudin, Terry Moore, and Stephen Levinson for discussion of the psychology and linguistics literature. Thanks also to audiences at talks in Chicago, Cambridge, and London.

[Originally published in *Mind*, 93 (1984), 161–76; with acknowledgements to the author and editor.]

myself to these three not because they exhaust the tenser's possibilities, or because other intuitions cannot be disposed of;[1] but because these three have a unified explanation which detensers have so far missed. The basis of the explanation is that both in observation and communication time-lags are usually short enough to be ignored. We can usually take observation to inform us of the present state of distant objects, and communication to inform us of the speaker's present belief. I argue this in Sections 2 and 3; I then explain the three intuitions in a way which does not make the present objective.

2. *Observing the present*

Observation takes time: the time-lag is due both to neural processing by the observer and, for sight, hearing, and smell, to the transmission of signals to our sense organs. We can however almost always ignore this time-lag. That is, if the process of observation is reliable, we can take ourselves to learn the observed object's state at the time we make our observational judgement, and not merely at some previous time. We can judge 'a is now G' or 'a is G at t' where t is the *time of judgement*; and not merely 'a was G' or 'a was G at t'', where t' is some previous time at which, we believe, signals left the object. Similarly for ascriptions of relations: we can judge 'a is now H to b' or 'a is H to b at t', though the time-lags involved in observing a and b may be rather different, say because a is further away than b.

That we can usually observe objects' present states has been remarked before (Dummett 1973: 388; Mellor 1981: 157). But without some estimates of the time-lags involved in observation the claim may be dismissed as mere conjecture (Zemach 1972: 253). (And I must admit that smell is an exception: we can smell burnt toast long after the toast has stopped burning. I only make the claim for the other senses: touch, sight, and hearing.) My estimates will concentrate on sight and on the ascription of properties rather than relations; the other cases can be dealt with similarly.

We must of course estimate the distances of observed objects, the transmission speeds of light signals, and the time we take to process these signals into observational judgements. Most objects we observe are less than 1,000 metres away, and light travels this sort of distance almost instantaneously. Since we process light entering our eyes in about half a second, that is the typical time-lag in observation.[2]

[1] For some examples, cf. Taylor (1955), Garson (1971), Mellor (1981: chs. 3 and 5), and Butterfield (1985).

[2] Processing takes about half a second for many kinds of observation; cf. Posner and Mitchell (1967), Potter and Faulconer (1975), Flowers (1975), and Shaffer and LaBerge (1979).

But we must also estimate how rapidly objects change their observable properties. If these properties changed fast enough, an object's closeness and the high speed and rapid processing of signals from it would not justify our judging it to have an observed property at the time of our judgement. In other words, we must estimate not only the time-lag, but also a limit the time-lag must not exceed if we are to be entitled to ignore it.

Estimating this limit needs some care. Call the shortest length of time for which *a* can be *G a*'s *time-scale* for *G*. Suppose for example it is five seconds. Prima facie then, a judgement '*a* is now *G*' (or '*a* is *G* at *t*', *t* the time of judgement) would need to be made within five seconds of the light leaving *a*, if it is to be correct.

But this is not quite right. We must allow *a*'s changing with respect to *G* within the time-lag of observation. Suppose *a* does in fact change with respect to *G* every five seconds, and that photons leading to an observation are equally likely to leave *a* at any time. Then if a judgement '*a* is now *G*' is made half a second after the photons which cause it leave *a*, its probability of being true will be 9/10. For photons leaving *a* in the first nine-tenths of a five-second period during which *a* is *G* (i.e. the first $4\frac{1}{2}$ seconds) will lead to true judgements; photons leaving in the last tenth (i.e. the last half-second) will lead to false judgements. (This of course assumes that our perception is reliable in other respects, i.e. that it will make us judge *a* to be *G* if it was *G* when the light left it.) Similarly for other time-scales: if a judgement '*a* is now *G*' is to have a 90 per cent probability of being true, the time-lag must be less than a tenth of the time-scale.

But how then can we claim to observe the distant present? We certainly think that fewer than 10 per cent of our observational judgements are falsified by objects changing during the time-lag in observation; yet given a time-lag of half a second, that means a time-scale of at least five seconds. And surely most objects can have most of their observable properties for less than five seconds.

Indeed they can: but they do not. In the above calculation I assumed that *a* changes with respect to *G* as often as it could change. That frenetic assumption is of course generally false, and its falsity allows us a longer time-lag for a given probability of truth. What we need is the *average* time for which *a* does not in fact change with respect to *G*, *not* the *shortest* time. Suppose for example that on average *a* changes with respect to *G* every ten seconds. Then with a time-lag of half a second, judgements '*a* is now *G*' get a probability of 95 per cent; and the less frequent the changes, the greater the probability will be.

This makes all the difference. Most of the observable objects around us are solids—'moderate-sized specimens of dry goods' as Austin called them

(1962: 8); and these actually change their observable properties very infrequently. So time-lags very rarely falsify our judgements about their observable properties at the time we make the observation; falsity in our judgement almost always arises for other reasons than a change in the object during the time-lag.

Given that we *can* usually ignore observational time-lags, *why* can we? Why is the time-lag about half a second; and why do most observable properties change rarely enough for that time-lag to be ignored? I think we should take the second question first: once we have explained why observable properties change as rarely as they do, we can expect there to be a functional (ultimately, evolutionary) explanation of why the time-lag is about half a second. In outline, the explanation would be that, provided perception is otherwise reliable, such a time-lag allows us confidently to make present-tensed, not just past-tensed, observational judgements. And that is useful since our actions are more likely to succeed if guided by true beliefs about the present, and not merely the past, state of the world; our actions may still fail, but at least we avoid failure due to changes occurring since the start of the process of perception.

Why then do most solid objects change their observable properties so rarely? Surely this is because of their resilience to the frequent internal and external disturbances of various kinds that bring about change, and the infrequency of major disturbances they would not be resilient to. I consider in turn mechanical, electromagnetic, chemical, and thermal disturbances. Most observable objects are too massive and too strong to be changed by common sorts of mechanical disturbance such as air pressure and low-speed impacts with other objects. They are nearly electrically neutral, and so not seriously subject to electromagnetic forces. They are resilient to most of the chemical reactions that could befall them, being chemically stable so that reactions will not originate internally, and massive enough to be little affected by most common chemical interactions with the environment. Finally, they are not often liable to be changed by internal thermal agitation, nor by external heat.

3. Communicating with the present

Communication takes time: in speech the time-lag is due to neural processing by speaker and hearer, the time taken to speak, and the time speech takes to reach the hearer. Similarly, of course, for written communication. For speech we can almost always ignore this time-lag: someone hearing the speaker assert that *p* can take himself to have learnt the speaker's present belief that *p* (supposing the speaker to be sincere), not just a past one. That is, he can take the speaker to believe that *p* at the *time of reception* of the message and not just at the *time of utterance*[3] (*p* need not of course be a present tense proposition; it can have any tense or be tenseless). As for assertions, so for requests; the hearer can take himself to learn what the speaker desires at the time of reception, not just what he desired at the time of utterance. And similarly for other speech-acts.

These claims are justified as in Section 2: the time-lags in communication are generally less than some allowable maximum. In observation the maximum is set by (i) some prescribed probability of correctly judging an object's state at the time of judgement and (ii) the frequency with which objects change their observable properties. Here it is set by (i) a prescribed probability of the hearer correctly judging what the speaker believes, desires, etc. at the time of reception, and (ii) the frequency with which speakers change the beliefs, desires, etc. that they communicate.

The time-lags in oral communication are generally longer than those in observation: usually several seconds—and mostly due to the time it takes to speak the utterance. Nevertheless they can usually be ignored. The reason, as in the case of observation, is that (ii) is so low that the time-lags stay below the maximum even if we set (i) high. This is clear from our experience. We habitually say what we believe or desire, confident that we will still believe or desire it when the hearer understands our utterance. There are exceptions (racing commentaries for instance!); but only in the much slower process of written communication is it at all common for us to change our minds before the message is received.

That time-lags in communication can usually be ignored greatly enhances an individual's ability to observe the distant present. When we hear an observational report '*a* is now *G*' or '*a* is *G* at *t*' (*t* is time of utterance) we can usually take the report to claim that *a* is *G* at the time of reception as well as at the time of utterance. And since other observers are usually sincere in their observational reports, and their perceptions as reliable as our own, we can usually believe that *a* is *G* at the time of reception. The point is obviously important when we have to coordinate with one another at a distance, e.g. getting a crane hook into the right position to attach the load, or moving a sofa upstairs.

Given that we can ignore time-lags in communication, one can still ask, as in Section 2, why this is so. Why is the time-lag typically several seconds? And why do the beliefs etc. that we communicate change rarely enough for

---

[3] Allowing for insincerity: the hearer can take the speaker to intend him to think that the speaker believes that *p* at the time of reception. I will ignore insincerity throughout; the amendments needed to allow for it are obvious. (Cf. how I ignored the unreliability of observation in Section 2.)

these time-lags to be ignored? My argument does not need answers to these questions, but I should emphasize that unlike the case of observation, we cannot expect to explain *first* why the beliefs etc. we communicate change rarely, and *then* give a functional (ultimately, evolutionary) explanation of why the time-lag is typically several seconds. A time-lag short enough to be ignored would indeed be useful just as in Section 2, since our actions are more likely to succeed if guided by true beliefs about the speaker's present, and not merely past, beliefs. But we cannot expect to treat the two questions separately. For it seems likely that the time-lag in communication influences which beliefs etc. we communicate. At least, such an influence, by for example limitations of the vocal tract, seems more likely than the parallel in observation: namely that the observational time-lag influences what counts as an observational property.

4. Restricting existence to the present

We are far more inclined to say that all and only present objects exist than that all and only objects that are at some time here exist. At first, it seems clear why there is this asymmetry. Both tensers and detensers admit that 'All and only present objects exist' is ambiguous; since objects are created and destroyed, we must distinguish two senses of existence. There is the narrow sense, present existence, in which Reagan exists and Caesar does not; and a wider sense, timeless existence, in which all past present and future objects exist.[4] Obviously, 'All and only present objects exist' is true or false, according as 'exist' means present or timeless existence. But in any case 'exist' never means 'are at some time here'; hence the asymmetry.

Detensers cannot however just stop there. Why does 'exist' never mean 'are at some time here', when it sometimes means 'are somewhere now'? It is surely no accident that the narrow sense of 'exists' is present existence and not its spatial analogue; and the reason for this asymmetry, whatever it is, may well undermine the detenser's analogy between the present and the local. Moreover, it is often suggested that present existence should be analysed by the existential quantifier; this means associating with each time a domain of quantification containing the objects that then presently exist, and using temporal operators to pass us from one time to another as in tense logic. But no one has made the analogous suggestion: that location here at some

---

[4] If talk about alternative possible futures or pasts requires objects that never in fact exist, there will be a still wider sense including such possibilia. Also, if one supposes there is no actual future one will reject timeless existence, though one will accept a wider sense including possibilia. But allowing for these complications will not affect the argument.

time should be analysed by the quantifier, with spatial operators used to pass us from one place to another. Why not?

Sections 2 and 3 above give the detenser answers to both these questions. He can explain why the narrow sense of 'exists' is present existence, but not being at some time here; *and* why present existence (but not being at some time here) can be analysed by the quantifier, *without* making the present objective. I consider these in turn.

That 'exists' means present existence follows from its use as an observational predicate which is ascribed to *whatever* can be observed, and which can apply to an object at one time and not at another. The first feature serves to contrast with 'is red' etc.; the only properties 'exists' will demand of its instances will be properties about location in space and time. The second feature serves simply to set aside tenseless existence, i.e. being at some time located somewhere. Since we can predicate 'exists' of distant objects, it does not mean 'is now (or was or is at some time) located here'. And since by Section 2 we can usually take observed objects not to be destroyed during the observational time-lag, 'exists' does not mean '*was* located somewhere' or '*was* located somewhere in my field of observation'. Rather it means 'is now located somewhere' or 'is now located somewhere in my field of observation'. This last possibility is ruled out by the fact that someone who is told that *a* exists will not refuse to assert that *a* exists merely because he himself cannot observe *a*.[5]

Sections 2 and 3 also explain why the quantifier can analyse present existence, but not location here at some time. We have only to assume that *ceteris paribus* we should analyse observational sentences (i.e. sentences reporting observational judgements) by simple rather than complicated formulae.

By Section 2, observation usually informs us of objects' properties and relations at one time (the time of judgement) though they are not then at one place. Thus observational sentences generally have a temporally local but spatially dispersed subject matter. The time can be specified by an adverbial expression governing the sentence; while spatial information will be naturally specified by adjectival expressions qualifying terms for the objects. So temporal operators give observational sentences simpler analyses than spatial operators do.

Furthermore, if the quantifier analyses present existence, present-tensed observation sentences can be analysed by formulae without temporal operators: '*a*

---

[5] This explanation carries over to other observational predicates. For example, we can similarly explain why 'is red' means 'is now red and is located somewhere'. I admit that some further anti-verificationist premiss is needed to prevent e.g. taking 'exists' ('is red') to mean 'is now (red and) located somewhere in someone's field of observation'. For a recent discussion, cf. Williams (1981).

is now *G*' and 'Something is now *H* to *b*' will be analysed by the formulae '*G*(*a*)' and '(∃*x*)*H*(*x*, *b*)'. But if the quantifier analyses location here at some time, i.e. we associate with each place a domain containing the objects that are ever there, such sentences will generally have to be analysed by formulae with spatial operators. For only with operators can we construct formulae mentioning objects in different domains. A sentence like 'Someone is taller than *b*' will thus get a complicated analysis, just as in tense logic 'Someone (now dead) was taller than *b*' gets a complicated analysis. Only if we wanted to give simple analyses to sentences with a spatially local and temporally dispersed subject matter, such as 'It often rains here,' would we use the quantifier to analyse location here at some time. But we surely want to do no such thing: such states of affairs as its often raining here seem to us intrinsically complex —because they do not reveal themselves in single observations.[6]

These considerations are strengthened by Section 3. For the rapidity of oral communication gives us pragmatic reasons for using observational sentences with a present tense like '*a* is now *G*', rather than those with a date like '*a* is *G* at *t*' or with another tense like '*a* was *G*'. If communication took long enough for *a* to have a good chance of changing with respect to *G*, or if we could not predict the time-lag, we might save our hearer having to guess or work out the time for which we are asserting *a* to be *G* by using a date: '*a* is *G* at *t*.' This is of course what we do by dating letters which take a long and unpredictable time to arrive. If the time-lag were long but predictable, we might use an indexical appropriate to the hearer's point of view: if we knew communication would take five minutes, we might say 'Five minutes ago, *a* was *G*.'[7] But as it is, oral communication is mostly fast enough for us to mean to be heard to say '*a* is now *G*,' and to be believed.

So much by way of explaining our tendency to prefer a concept of present to one of spatially local existence, and to analyse it by the existential quantifier. I must emphasize that in arguing against a quantifier analysis of location here at some time, and against spatial operators, I am not arguing for the quantifier analysis of present existence, nor for temporal operators— both commonly advocated by tensers.

First, we can give present-tensed observational sentences simple analyses *without* using the quantifier to analyse present existence, or using temporal operators. Wilson (1955), for instance, suggests placing all past, present, and

---

[6] This argument against spatial operators does not require them to exclude temporal operators. Even if we took an interpretation to represent an instantaneous state of the world, and had temporal operators passing us between these states as in tense logic, the subject matter of formulae without operators would still be spatially local.

[7] Cf. how a Latin letter-writer may use a past tense to describe events contemporaneous with his act of writing, cf. Cicero, *Fam.* 5. 12. 2.

future objects in a single domain, together with times which can be referred to by dates or by indexicals like 'now'. Present existence is thus represented by a two-place predicate: 'Exists (*a*, now).' Tenses are expressed by using a *later-than* predicate, quantifying over times and giving other predicates an extra argument-place for a time; thus '*a* was *G*' is analysed as '($\exists t$) (*Later* (*now*, *t*) & *G*(*a*, *t*))'. On Wilson's view, present-tensed observational sentences get simple analyses: they have no quantification on the argument-place for times, which is filled with 'now'.[8]

Secondly, there are obvious difficulties in arguing from Sections 2 and 3 for the quantifier analysis of present existence, or for temporal operators. I will consider here only the case of the quantifier; my (1984) considers temporal operators. Even if we were willing to identify existence (in the sense analysed by the quantifier) with some kind of observability, Sections 2 and 3 would not establish that all and only presently existing objects are in some sense observable. Section 2 shows only that *most* of the objects which we each observe presently exist, since most will not have been destroyed during the observational time-lag. A few may well have been destroyed, and even if they have all survived, none of us can observe *all* presently existing objects. Indeed, some presently existing objects are observed by nobody. So even if we pool our observations—i.e. take the quantifier to analyse present observability by some observer or other—not every presently existing object will occur in the domain of quantification.

5. *Time-variable and space-variable truth-value*

As Dummett says (1973: 390), 'the notion of truth-value as changing from time to time has a much more powerful appeal to us than that of truth-value as changing from place to place'. Why? After all, time-variable and space-variable truth-values are apparently analogous: both can be introduced by an indexical term ('last year's boat race', 'the bridge north of here') or by an indexical predicate ('was a fiasco', 'is five miles away'). So why does time-variable truth-value occur more often, or more prominently, than space-variable truth-value?

I think there are two reasons. The first is language-dependent and has nothing to do with Sections 2 and 3. It is, roughly, that the grammars of English and many other languages often make time-variable but not space-variable

---

[8] Quine's better-known (1960: section 36) analysis is similar. But the difference—that for Quine there are terms like '*a-at-t*' standing for *temporal parts* of objects—means that Quine cannot analyse relations between simultaneous temporal parts any more simply than those between non-simultaneous ones. For different terms in a sentence can have different temporal suffixes, e.g. '*H*(*a-at-t, b-at-t'*)'.

truth-value compulsory. Thus, apart from a few cases like 'A stitch in time saves nine' and '2 and 2 is 4', verbs in English are tensed: they convey information about the relation of the time spoken about to the time of utterance. (Similarly for other Indo-European languages.) Since almost all sentences contain verbs, most of them have a time-variable truth-value. On the other hand, there is no rule that English verbs, or sentences, must convey information about the spatial relation of the objects mentioned to the place of utterance. So unless they happen to contain a spatially indexical term ('the bridge north of here') or predicate ('is five miles away'), they will not have a space-variable truth-value.

This explanation does not of course depend on *verbal* tensing—though some detensers' remarks suggest it does (e.g. Quine 1976: 147). It applies to any language in which most sentences relate the time spoken about to the time of utterance, but not the place spoken about to the place of utterance. Whether this information is conveyed as in English, by verbal inflections and auxiliaries, or by adverbs like 'soon', 'now', and 'recently', or by adverbial phrases like 'in a while' and 'five days ago', is quite immaterial.[9] But the explanation will not apply to languages in which spatial indexicals are at least as frequent as temporal indexicals. The Amerindian language Kwakiutl is like this. Kwakiutl verbs are indeed tensed; but its nouns also have compulsory prefixes which convey whether the object mentioned is near the speaker, or near the audience or near some other person spoken of, and whether it is visible or invisible. (These prefixes also serve as pronouns: thus the third-person pronoun 'he' has 6 (= 3 × 2) basic forms to represent who *he* is near and whether *he* is visible; cf. Boas 1911: 445–6, 527–9.) So Kwakiutl speakers might find space-variable truth-value as 'appealing' as time-variable truth-value.

But they might not. For there is another reason for the greater appeal of time-variable truth-value, which is quite independent of language. It relates to observational reports of the kind we can make before learning the use of watches, maps, etc. In making these reports, we can much more readily specify the objects reported on without using spatial indexicals then we can specify non-indexically (e.g. without reference to 'now') when they have the reported properties and relations. Thus we can often avoid using spatial indexicals, but we rarely avoid using a temporal indexical like 'now' or a verbal present tense.

This claim does not of course conflict with the existence of a language like Kwakiutl. That we can often specify an object without spatial indexicals

---

[9] Chinese and Finnish, which are sometimes cited as tenseless languages, in fact have some verbal tensing (cf. Chao 1948: 54–5; Olli 1958: 11, 25 ff., 79 ff.). So far as I can judge, most Chinese and Finnish sentences relate the time spoken about to the time of utterance (by some combination of auxiliaries and adverbs), but do not relate the objects mentioned to the place of utterance; so this explanation applies to them.

does not mean that we have no spatially indexical information about it, e.g. whether it is near the speaker or the audience. We do have such information, and can incorporate it into our observational report if our language, like Kwakiutl, requires us to do so. The claim would only conflict with a language which required observational reports to specify non-indexically the time for which a predicate is ascribed. For it suggests that unless speakers of such a language were supplied with watches etc. they would often be tongue-tied, too ignorant to speak—a hardly believable state of affairs. At any rate, I know of no such language.

The reason why we can usually avoid spatial indexicals is that there is usually enough qualitative variation across space. That is, we can usually distinguish the objects we mention, at least from other currently observable objects, by their qualities—or at least by their qualities and their non-indexical spatial relation to other objects so distinguished.

On the other hand, watches etc. aside, we usually cannot specify the time non-indexically because the natural 'watches' provided by daily and seasonal variations are too crude. Using them we can only describe the present time as, for example, during early morning on a summer day; and that is imprecise compared with an indexical like 'now'. The latitude allowed by such descriptions is at least an hour—far greater than the period usually required to make an observation and communicate it to one's audience, i.e. a few seconds. So such descriptions cannot generally be substituted in observational reports for an indexical 'now', or a present tense, without a crucial loss of information—crucial because the observed objects will like as not change in the course of an hour.

Of course, the world might have had much less qualitative variation across space, and might have had natural 'watches' by which we could describe the present time with a latitude small enough for the descriptions to be useful in observational reports. One can imagine a world with objects, changing colour continuously and in step with one another, and common enough for everyone to be able to see one. But although this is imaginable, there are, as we have already seen in Section 2, physical reasons for not expecting such frequent changes in objects' observable properties. So though this reason for the prevalence of temporal indexicals and thus of time-variable truth-value is contingent, the contingency is not surprising.[10]

---

[10] I admit that to get the idea across, I have expressed this reason in a language-dependent way. I have talked of specifying *objects* with spatial indexicals and *times* with temporal indexicals; and this shows a bias against temporally indexical specifications of objects and spatially indexical predicates. To put the point in a language-independent way, think of a report as requiring a property of some spatial region for some interval of time; the point is then that we can more often specify the region non-indexically than the interval.

6. *Sharing a now but not a here*

We tend to think that we share a common, albeit ever-changing, *now* while we each have a different *here*. Since we generally want some connection between the objectivity of a notion and different individuals' agreement on it, this intuition apparently supports the tenser's claim that the present, but not the local, is objective. However, the detenser can explain this intuition by using Sections 2 and 3 to explicate the metaphorical ideas of sharing a *now*, and sharing a *here*.

The rapidity of oral communication (Section 3) suggests that people share a *now* but not a *here*, in that their interpretations of tokens of 'now' but not 'here' agree. That is, each person can usually take a token of 'now' which they hear to refer to the time of reception, as well as the time of utterance, without misinterpreting the speaker. But they cannot analogously take a token of 'here' to refer to where they themselves are.

Explicated like this, a detenser can obviously endorse the intuition. But is this all there is to it? Suppose that communication were much slower than it actually is; suppose for example that we communicated entirely by letter and not at all by speech, gestures, etc. Wouldn't we still have the sense of sharing a now but not a here? I think we would; but nevertheless the detenser can still explain the intuition, though now in terms of observation rather than communication. But this explanation is not an endorsement: if one explicates sharing a now or here in terms of observation, then people share a here in strict analogy with their sharing a now. Rather the detenser explains the intuition by explaining why we tend to see only one half of this analogy, namely the sense in which people share a now.

The obvious way to explicate a common now without appeal to communication is to say that two people share a now if they agree in their judgements about what is now the case, i.e. in their present-tensed judgements. However this suggestion needs clarifying on two counts; and doing this will lead us to consider observational judgements. First, if present-tensed judgements are made at different times they cannot be expected to agree. So we must apparently fix the reference of the 'now', and explicate sharing a now *at a time*. Secondly, fixing the now should not be sufficient for agreement; two people can surely share a now while having disagreements, provided these disagreements can be explained by their different theoretical beliefs, past experiences, spatial perspectives, etc. Presumably these factors are set aside in the case of ascriptions of observational predicates, other than spatially indexical predicates like 'is far away', to objects that both people can observe and that are not specified in a spatially indexical way.

Putting these two points together, let us say that two people share a now at a time if they then make the same ascriptions of such predicates to such objects. Section 2 shows that people with reliable perceptions usually do share a now in this sense, for different people's time-lags in the process of arriving at such ascriptions are roughly equal. But this is of course a contingent fact; the time-lag might have been very different for different people, say because some people were as slow-witted in their observational judgements as dinosaurs—in which case we would not share a now in this sense.

This explication of sharing a now involves agreement of simultaneous observational judgements without spatial indexicals about what is now the case. The analogous explication of sharing a here therefore requires agreement of observational judgements made at the same place without temporal indexicals, about what is here the case. And on this explication, people do share a here at a given place, in strict analogy with their sharing a now at a given time: two people with reliable perceptions will make the same observational judgements of the form '*a* is here *G* at time *t*' provided they make them at the same place. (This again is a contingent fact. If people were slower-witted in their judgements about where they are, they would sometimes use 'here' to refer to a place they had been rather than to where they were; and so judgements made in the same place about what is here the case would sometimes disagree.) So the detenser must explain why we do not see this analogy but instead think of ourselves as sharing a now but not a here.

Fortunately, the preceding discussion suggests three reasons for this; the first two come from Section 5, and the third from Section 4. And the detenser can rely on any or all of them, as he likes; I believe all three. First, since our language is tensed we take 'sharing a here' to mean *now* sharing a here. And people do not usually *now* share a here, if we explicate this on analogy with their now sharing a now, i.e. if we require agreement in their simultaneous judgements about what is here. That is, let us say that two people share a here at a time if they then make the same observational judgements about what is here. (As before, we consider only observational judgements, so as to set aside disagreements due to different theoretical beliefs, past experiences, etc.) In this sense, people do not usually share a here; even if they are close enough to observe the objects located at each other, they are usually far enough apart for these to be two different objects, with in general different properties.

Also, the sense in which we share a here (at a given place) is recherché, for two reasons. First, this sense involves observational judgements like '*a* is here *G* at *t*' that have spatial but not temporal indexicals. And once we set aside devices like watches, we can avoid temporal indexicals in this way much less often than we can avoid spatial indexicals, as required in the judgements '*a* is now *G*' etc. associated with our sharing a now. Second, the sense in which we share a here is recherché on account of Section 4's point that the rapidity of oral communication gives us pragmatic reasons for making present-tensed observational reports. Only if the time-lags in oral communication were much longer or more variable would we follow our practice with letters and use a watch etc. to specify non-indexically the time spoken about; and in that case, judgements like '*a* is here *G* at *t*' would not be as recherché as they are.

REFERENCES

Austin, J. (1962), *Sense and Sensibilia* (Oxford: Clarendon Press).

Boas, F. (1911), *Handbook of American Indian Languages*, vol. i (Washington: Smithsonian).

Butterfield, J. (1984), 'Dummett on Temporal Operators', *Philosophical Quarterly*, 34: 31–42.

—— (1985), 'Spatial and Temporal Parts', *Philosophical Quarterly*, 35: 32–44.

Chao, Y. (1948), *A Mandarin Primer* (Cambridge, Mass.: Harvard University Press).

Dummett, M. (1960), 'A Defence of McTaggart's Proof of the Unreality of Time', *Philosophical Review*, 69: 497–504.

—— (1968), 'The Reality of the Past', *Proceedings of the Aristotelian Society*, 69: 239–58.

—— (1973), *Frege: Philosophy of Language* (London: Duckworth).

Flowers, J. (1975), 'Sensory Interference in a Word-Color Watching Task', *Perception and Psychophysics*, 18: 37–43.

Gale, R. (1964), 'Is it now now?', *Mind*, 73: 97–105.

—— (1968), *The Language of Time* (London: Routledge & Kegan Paul).

Garson, J. (1971), 'Here and Now', in E. Freeman and W. Sellars (eds.), *Basic Issues in the Philosophy of Time* (La Salle: Open Court): 145–54. Also in *Monist*, 53 (1969): 469–77.

Lloyd, G. (1978), 'Time and Existence', *Philosophy*, 53: 215–28.

Mellor, D. (1981), *Real Time* (Cambridge: Cambridge University Press).

Olli, J. (1958), *Fundamentals of Finnish Grammar* (New York: Northland Press).

Posner, M., and Mitchell, R. (1967), 'Chronometric Analysis of Classification', *Psychological Review*, 74: 392–409.

Potter, M., and Faulconer, B. (1975), 'Time to Understand Pictures and Words', *Nature*, 253: 437–8.

Prior, A. (1959), 'Thank Goodness That's Over', *Philosophy*, 34: 12–17.

—— (1970), 'The Notion of the Present', in J. Fraser (ed.), *The Study of Time* (Berlin: Springer).

Putnam, H. (1967), 'Time and Physical Geometry', *Journal of Philosophy*, 64: 240–7.

Quine, W. (1960), *Word and Object* (Cambridge, Mass.: MIT).

—— (1976), *Ways of Paradox* (Cambridge, Mass.: Harvard University Press).

Shaffer, W., and LaBerge, D. (1979), 'Automatic Semantic Processing of Unattended Words', *Journal of Verbal Learning and Verbal Behaviour*, 18: 413–26.

Taylor, R. (1955), 'Spatial and Temporal Analogies and the Concept of Identity', *Journal of Philosophy*, 52: 599 612.

Williams, B. (1981), 'Another Time, Another Place, Another Person', in *Moral Luck* (Cambridge: Cambridge University Press): 164–73.

Wilson, N. (1955), 'Space, Time and Individuals', *Journal of Philosophy*, 52: 589–98.

Zemach, E. (1972), '"Here" and "Now"', *Mind*, 81: 251–5.

4

Tense and Emotion

DAVID COCKBURN

I

In so far as the mind conceives a thing under the dictates of reason, it is affected equally, whether the idea be of a thing future, past, or present. (Spinoza 1955: IV. lxii)

One of the strands of argument by which Spinoza reaches this conclusion starts from the following thought:

So long as a man is affected by the image of anything, he will regard that thing as present, even though it be non-existent . . . he will not conceive it as past or future, except in so far as its image is joined to the image of time past or future . . . Wherefore, the image of a thing, regarded in itself alone, is identical, whether it be referred to time past, time future, or time present. (1955: III. xviii, proof)

There is a core in common to the belief that it *has* rained, the belief that it *will* rain, and the belief that it *is* raining. That common core is to be seen in its pure form in the last case. It becomes a belief that it *has* rained, or a belief that it *will* rain, by undergoing a certain modification: by being 'joined to the image of time past or future'.

This account of the relation between the beliefs that something *is* happening, *has* happened, or *will* happen is linked by Spinoza with an account of the relation between the emotions that we feel in the three cases. Just as there is a common core to the beliefs that a thing is past, present, or future, so there is a common core to the emotions that we feel in the face of a past, present, or future event of a certain kind. But that common core, seen in its pure form in our response to something we believe to be happening now, is modified by the uncertainty that we almost always feel about the occurrence of a past or future event. The pain that I feel on being confronted with some distressing event is, when the event is anticipated rather than present, transformed into fear: 'an inconstant pain also arising from the image of something concerning which we are in doubt' (1955: III. xviii, note ii).

I would like to thank Robin Le Poidevin and an anonymous reader for Oxford University Press for helpful comments on an earlier draft of this chapter.

Now in the deterministic world assumed by Spinoza the clear-thinking and well-informed individual will feel no uncertainty about past or future events. He will see that the past evil *has* happened, or that the future evil *will* happen, with the same clarity as he sees that the present evil with which he is confronted *is* happening. The lack of constancy which marks off my response to an anticipated evil from my response to a present evil will thus disappear, leaving my feelings about the two evils identical: 'In so far as the mind conceives a thing under the dictates of reason, it is affected equally, whether the idea be of a thing future, past, or present.'

The success of Spinoza's argument turns in part on the correctness of his account of why we do, in practice, feel differently about past, present, and future goods and evils. Now it seems fairly clear that no account along the lines he offers, that is to say exclusively in terms of our doubt or certainty about the occurrence of an event, has much chance of success. While part of the reason I care less about my own distant future than about my immediate future may lie in the uncertainty which surrounds it, is it not clear that we often choose the immediate pleasure even when we are in no doubt that it will be at the cost of greater suffering in the future? We take the extra drink knowing that it will lead to a hangover tomorrow. We eat, drink, and smoke too much knowing we will pay a price in twenty years' time: a price that we would never dream of paying if we had to put up with the discomfort now for the pleasures in twenty years' time.

Spinoza might reply that our 'short-sighted' behaviour shows that we do not really accept that we will have to undergo these things in the future. We may mouth the words 'I will suffer terribly tomorrow' as we down another whisky; but mouthing the words is not at all the same thing as really believing them in the sense that is relevant to our actions and feelings. Failures of prudence always reflect a failure fully to grasp, in a deep sense, the implications of one's actions.

More generally, it might be argued that there is a clear sense in which our emotions track the vividness of our conception of events in a way which is close to what Spinoza has in mind. When the dangers of smoking are first vividly portrayed to me I am filled with fear and manage to give it up for a week or two. But I gradually assimilate the idea; even though I still 'know' the dangers, in the sense that I can recite the statistics, the thought of them gradually loses its grip on my imagination and I slip back, with no more than mild unease, into my old ways. Again, when I first learn that I have failed a crucial exam, or first realize that I have behaved in a disgraceful, or in an appallingly embarrassing, way, I feel terrible and have trouble facing the world. Now while my *certainty* about what happened does not weaken with time, the imagery of the incidents nags at me less insistently and loses its initial vividness. And as the idea in this way becomes familiar, the associated emotions lose their force. Similarly with the grief that I feel when I first learn of the death of a loved one. The grief softens with time: as a result, not of any weakening in my certainty that she died, but of the way in which the idea of her as dead becomes familiar to me; or, in another case, as a result of the way in which other things start to fill my life and so she becomes less important to me.

Now if the pattern in our emotional life is dictated in this way by, on the one hand, the shock of the unfamiliar and, on the other, a dulling of appreciation with time, will there not be room for an argument very close to Spinoza's? A being whose awareness of past and future was of the same adequacy and vividness as her awareness of what is happening now in her immediate vicinity would not be liable to such fluctuations in her emotional life. What is happening *now* would hold a no more central and vivid place in her thought than what *will* happen next week or what *did* happen twenty years ago. Everything that happens, whenever it happens, would be to her equally familiar and assimilated; and so her feelings about an event would not vary as it moves from the future, through the present, into the distant past.

Suggestions of this form are often linked with the idea that one who sees things clearly will view everything which happens with equanimity. Thus, Spinoza suggests that the strong man will appreciate that what 'seems to him impious, horrible, unjust, and base, assumes that appearance owing to his own disordered, fragmentary, and confused view of the universe' (1955: IV. lxxiii, note). A further argument will be needed, however, to reach that conclusion. It does not follow from the fact that I will respond in the same way to my recognition that my child will die some day, that my child has just died, and that my child died twenty years ago, that my response will be characterized by the calmness with which I can, perhaps, think of my child's death in the last case. So far as this argument goes it is possible that the clear-thinking man will have a steady vision of his child's death as an unmitigated horror. Be that as it may, can this modified version of Spinoza show that our failure to be moved in the *same* way by what happens in the past, present, and future is the product of ignorance, lack of imagination, or irrationality of some form?

I doubt it. Consider first the man on his seventeenth whisky who, it is suggested, is merely mouthing the words 'I will suffer terribly tomorrow.' It seems likely that this charge will have to be *based on* the conviction that his behaviour is irrational. We simply refuse to count as 'seriously believing that if he carries on like this he will suffer terribly tomorrow' the man who nevertheless reaches for another drink. Now if it is like that, then the suggestion that those who are 'imprudent' have not fully grasped the consequences of their behaviour cannot be used to support the charge of irrationality. Someone who is inclined to say that it is quite appropriate to abandon oneself totally to the immediate pleasure will simply insist that a failure to be concerned about tomorrow's hangover does not at all show that I have not fully grasped that it really will happen.

An analogous point applies to the suggestion that, for example, the difference between the intensity of my grief when a loved one has just died and what I feel ten years later is to be explained in terms of some change in me to which a fully adequate being would not be subject. The argument assumes that if my grief softens with time we are compelled to speak in terms of a distortion in the place which the death occupies in my thought, either when it first occurs or ten years later. But if, for example, we insist that the softening of my grief must reflect a 'dulling' of my awareness of what happened we are simply assuming what is supposed to be proved: that a softening of grief is not a completely appropriate development as a death recedes into the past.[1]

II

It seems likely, however, that the philosophical demand for equal concern for all times characteristically arises ultimately from a metaphysical picture of time which lies at a deeper level than anything we have yet touched on. The crucial idea might be expressed in the words: 'All times, and hence all the individuals occupying them, are equally real' (Le Poidevin 1991: 5). Our placing of events in the categories 'past', 'present', and 'future' does not distinguish them in terms of some deep division in reality: does not distinguish them in terms of some difference between the events themselves. Thus, when some event—say the suffering or death of a loved one—'moves from the future into the present' this is not a matter of its acquiring some new characteristic: as the poker acquires a new characteristic as it changes from being cold to being hot. The role of, for example, the future tense is not to indicate some *property* of the event spoken of. The tense of an utterance serves merely to indicate the temporal relation in which the speaker stands to the event: to indicate, for example, that the event is later than the time of utterance. As the point is sometimes put: while there are tensed judgements there are no tensed facts. The truth-conditions of my judgement that it will rain

---

[1] Of course, the phrase 'recedes into the past' belongs with a way of thinking of time which is the central target of those whose views I am discussing in this chapter. But I am not here begging the question in favour of that way of thinking; just insisting that the question must not be begged against it.

tomorrow—that objective feature of the world which makes my judgement true—is the fact that it rains on the day after I make the judgement.[2]

To express the point in terms which I used in my presentation of Spinoza, there is a common core which is expressed by the assertions 'It has rained,' 'It is raining,' and 'It will rain'; and the tense of the utterance serves merely to indicate where this common core—'rain falling'—stands in relation to the utterance: as the words '100 yards to the east' serve merely to indicate the relation in which an event stands to the speaker. Thomas Nagel expresses the point in this way:

the sense of an assertion about what is the case at a certain time does not change with the tense of the assertion; the tense merely indicates a relation between the time of utterance and the time of what is being talked about.[3]

It is a short step, if a step at all, from this claim to the conclusion that all past, present, and future events are, in themselves, of equal significance. We could compare here the fact that a particular pain in my leg is very intense with the fact that it is happening *now*. The first fact might clearly be regarded as conferring a special status on that pain. It gives me a reason to be more concerned about it than about the milder pain in my hand. But the second 'fact', the fact that the pain is happening *now*, seems to be in a very different position. To say that the pain is happening now is not to ascribe some further property to the pain; the role of the word 'now' is simply to indicate that the pain is occurring at the same time as the utterance of the words. And that, presumably, cannot be regarded as conferring any special status on the pain considered in itself; any more than the fact that a man is suffering 'here' confers any special status on the suffering considered in itself. In itself, then, a future pain is as much something to be avoided as is a present pain; and a failure to acknowledge this in one's behaviour reflects a failure of rationality.

My presentation of this argument is modelled primarily on part ii of Thomas Nagel's *The Possibility of Altruism*. Despite the close links between his argument and that of Spinoza, Nagel's understanding of what he is doing is rather different from Spinoza's. Spinoza thinks of himself as calling for radical revisions in our emotional and practical lives. Nagel, by contrast, takes himself to be spelling out the rationale for ways in which we already live and think. We are in practice concerned about what happens at times other than the present. We are, for example, moved by the prospect of our own future suffering; and even when we fail to give much weight to it, as with the anticipated

---

[2] The 'tenseless theory', in one form or another, has dominated recent philosophical treatments of time. My presentation of the view owes most, perhaps, to Mellor (1981).

[3] Nagel (1970: 61). See also Ayer (1954: 186).

hangover, we have an intuition that the future event *should* have an effect on our present actions. Nagel presents his argument as a defence of the thesis that 'such intuitions are objective and possess an explanation which supports their validity' (1970: 57). It is, for example, not simply that we are beings of a kind who *happen* to be concerned about our future well-being. We have reason to be concerned. That something lies in the present does not give it any special claim on my actions, for 'there is reason to *promote* that for which there is *or will be* a reason' (1970: 45).

It might, however, seem clear that any argument along the lines proposed by Nagel or Spinoza is, as Spinoza insists, bound to be strongly revisionary. For is there not a clear sense in which virtually all of us do, in practice, give much greater weight to the present and immediate future than to the distant past and future? And, further, would not the great majority insist that it is, in many ways, quite in order that we should do so? Nagel claims that 'the fact that a particular stage [of one's life] is *present* cannot be regarded as conferring on it any special status' (1970: 60). Would not a full assimilation of this claim involve, as Spinoza suggests, radical changes in our emotional and practical life?

Nagel argues that it would not. A first point to note here is that it does not follow from the insistence that reasons are timeless in this way that thoughts about his own past, present, and future will influence the rational being's *behaviour* in the same way. While the fact that a certain event lies in the past makes no difference to its intrinsic significance, it does, as a consequence of the direction of causation, make a difference to what I can *do* about it. Similarly, while the demands that my future welfare make on me now are exactly the same as those which any present welfare makes on me, and so, other things being equal, should occupy the same position in my thought and behaviour, in practice other things rarely are equal. For 'the simple uncertainty of future needs, desires, and circumstances and the constant possibility of death provide additional arguments against excessive planning, despite the fact that reasons are timeless' (1970: 73).

None of this involves any withdrawal from the claim that past and future events have, in themselves, exactly the same importance as do present ones. It is simply to acknowledge the practical limitations on what we can *do* about the past and distant future. But what about how we should *feel* about past and future events? Will not an acceptance of the equal significance of past, present, and future events involve a radical change in our emotional life? Recent discussions of the relation between a tenseless view of time and our concern about the past have been dominated by an example presented by A. N. Prior in his paper 'Thank Goodness That's Over'. Prior's explicit target is the suggestion that tensed statements—for example, my words 'The pain is over'

—can be translated into tenseless ones. That, he argues, cannot be correct, for my sigh of relief 'Thank goodness that's over!'

certainly doesn't mean the same as, e.g. 'Thank goodness that date of the conclusion of that thing is Friday, June 15, 1954', even if it be said then. (Nor for that matter, does it mean 'Thank goodness the conclusion of that thing is contemporaneous with this utterance'. Why should anyone thank goodness for that?) (Prior 1959: 17)

Now recent defenders of a tenseless view of time accept that no such translation is possible. Their claim is rather that there are no tensed facts: in particular, no fact of my pain being over; or perhaps, in a weaker version, that if there is such a fact it is not one which is of any significance in itself. Prior's example, however, also poses a challenge to this version of the thesis. For we can ask: if there are no such facts just what is it that I am thanking goodness for? And, following Prior, we can add that the state of affairs which, according to the tenseless view, makes true my judgement that the pain is over is not, on the face of it, at all the kind of thing about which it would make sense to feel relief. (I must confess that I am unsure how one might take the argument further if someone simply denied this claim. Perhaps one might reasonably ask that some attempt be made to make the denial credible. In any case, as we will see, the most familiar responses from tenseless theorists are rather different.)

Nagel's response to Prior's example is to acknowledge the existence of such feelings of relief, but to add: 'insofar as this feeling provides reasons for action, those reasons will be timeless: viz. reasons to accept inevitable suffering without too much delay, thereby shortening the pains of anticipation' (Nagel 1970: 71). It is not, however, clear how this is supposed to help. For if we accept that 'reasons are timeless' we must surely accept that for the clear-thinking individual the pains of anticipation will be no worse than the pains involved in looking back on past suffering. What I am feeling relief about is 'the fact that the pain is *past*'. Since there is no such fact my relief is misplaced.[4]

It is, I believe, important when considering such points to bear in mind that there is not just one kind of case which creates difficulties for Nagel's view: not just one way—that seen in the case of our concerns about physical pain—in which *when* something happens makes a difference to our feelings about it; and not just one way in which the tense of an assertion makes a difference to how it can feature in the giving of reasons for action and feeling. Consider an example which Nagel himself discusses:

[4] This appears to be the implication of the view taken by MacBeath in MacBeath (1983).

timeless reasons explain the phenomenon of wanting some thing *to have happened*, simply because there was reason for it in the past. Someone who was too drunk at last night's party to remember what happened will hope that he behaved with restraint —not only because of the subsequent disagreeable effects of idiotic behaviour, but because there was reason to avoid it *then*. (1970: 72)

Despite the strong future orientation of much contemporary moral philosophy the past does, in practice, have enormous significance for us in many ways. In so far as Nagel's aim is to show the rationale for our current ways of thinking, it is no objection to his account that it implies that what *has* happened should sometimes matter as much to us as does what *is* happening or what *will* happen. For it is an obvious feature of our normal thought that, for example, 'I behaved idiotically' can intelligibly be given as a reason for my feeling terrible today.

Having said that, it must be added that there would be serious problems with his account if it should turn out that the *particular ways* in which past, present, and future happenings matter to us are often very different. For example, consider again last night's party. If I am slightly drunk at a party I may hope that I behave with restraint. This hope will find expression in my struggling to do so. After the party I hope that I did behave with restraint. My hope at this time cannot find expression in *that* way. We might be tempted to think of this, as Nagel does, as in no way casting doubt on the suggestion that what is felt, 'hope', is of just the same character in the two cases. The 'hope' felt later is simply expressed differently as a result of my different causal relationship to the event in question. If we broaden our view, however, this way of representing the situation becomes impossible. For example, if I discover that I did behave idiotically I may think that I have reason to apologize to the host and to avoid parties with the same group of people for a month or two. We can add that I do not simply *want* it to be the case that I behaved with restraint. If I did not, I *regret* that I did not. I may feel ashamed. Now in what sense is all of this a reflection of anything that I had reason to do or feel at the time of the party? (Might I not, at the time of the party, feel shame about what I am doing? Yes. But that leaves the question: is the relation between shame about what I am *now* doing and shame about what I *have* done of the form required by Nagel's argument? In particular, is the shame that I might feel at the time any more basically and transparently in place than the shame that I feel later?)

Consider another example. I hear that a loved one, who, let us suppose, emigrated some years ago, has died. I knew that he would die some day. Perhaps I assumed that he would die before I did, and that I would never see him again. I did not have hopes, dashed by the news of his death, that, for example, his life would be marked by some crowning achievement before he died.

Yet I feel deep grief on hearing that he *has* died. I cannot stand the thought that he is no longer alive. (It is, perhaps, important, when thinking about this case, to think also of a case in which someone knows the particular time at which a loved one will die: a case, for example, in which a loved one is to be executed.)

As a final illustration here, consider the kind of importance which the past of a person or of a place may have for us. Consider the way in which the words 'This is my daughter' or 'This is the man who committed the murder' may feature in the explanation or justification of my feelings towards, or treatment of, a person. Or consider the importance which the fact that a house has a history may have for our feelings about it. The thought that people have lived here—have sat reading on this bench, have played with their children in this hall, have had dinner parties in this room—perhaps over many generations, gives, for many, a richness to a house which may be fundamental to their feelings for it. Further, that added richness may not be compromised by the marks of wear and tear; on the contrary, the worn floorboards, the slightly musty smell, and so on may be tangible embodiments of the past which they would on no account sacrifice for the sparkling newness of a modern bungalow. Similar considerations apply in connection with the idea that the town in which one lives has a past. The fact that people have walked down these very streets for hundreds of years, or, in a rather different way, the fact that this is the spot at which many have been hanged, may transform one's feelings for a town. Now the thought that many others *will* live in a brand new, purpose-built house or city can, no doubt, be moving too; but it is not the same.

Nagel's problem here is to show how anything like our current emotional life might be consistent with the claim that 'past', 'present', and 'future' events all have exactly the same kind of reality, and so, in themselves, are all of exactly the same significance. His problem, that is, is to show how familiar ways in which we offer 'the fact' that something *has* happened, *is* happening, or *will* happen as a reason for actions and feelings can be acceptable if 'there are no tensed facts'. Now it might be thought that the practical considerations mentioned earlier—those bound up with the fact that we have control over the future but not the past—could be developed further in a way which might bear on our emotional life. It might be argued, for example, that emotional anxiety about future pains serves a practical purpose: it leads us to seek ways to reduce, or completely remove, the pain that will be felt. By contrast, given that our actions cannot influence what happens earlier than them, an emotional absorption in a past pain could serve little practical purpose.[5]

[5] See Seddon (1987: 88).

The suggestion here might be simply that we can *explain* a piece of irrationality in our emotional life: namely, our greater concern about future than about past pains. Alternatively, it might be suggested that these practical considerations explain the asymmetries in a way that justifies them. But whether the aim is simply to explain, or to explain in a way that justifies, there is room for serious doubt as to whether anything along these lines could account for our normal ways of thinking. There are problems about future suffering about which we do not believe that we can do anything—I take it that, in practice, people as readily fear inevitable pain as they do pain which they think might still be avoided; and I suspect that nobody but the most die-hard tenseless theorist will be tempted to believe that such fear is irrational. Problems arise too from those aspects of the past—on which I have briefly touched—about which we *do* care, sometimes intensely. If we are to be left with anything like our normal ways of thinking, we will need a supplementary explanation of why, despite their apparent lack of practical benefits, remorse and shame are such marked features of human life. (Perhaps they lead us to behave better in the future. But then perhaps a greater horror at past pains than we now have would lead us to avoid pains more effectively in future.)

The problem, then, is not simply that there are marked asymmetries in our normal concerns about past and future happenings; there is the further complication that the asymmetries take very different forms in the case of different kinds of happening. Now a quite detailed argument would be needed to show that no explanation of these asymmetries along the general lines suggested above could possibly be successful, and I do not know whether such an argument might be available. What is of much greater concern to me than this, however, is the attitude towards emotion which is reflected in recent discussions of these issues; and, in particular, the attitude towards emotion which is reflected in the idea that such an argument might justify the current form of our emotional life. It is being suggested that what makes it appropriate that we feel quite differently about, for example, past and future pains is the fact that this asymmetry of concern has practical benefits. Now the suggestion here might be that the only criterion of appropriateness for a form of emotional response is the benefits which such a response brings; or it might be, if this is different, that, while an emotion can be assessed in terms of its *appropriateness* to its object, practical considerations always overrule considerations of appropriateness in determining what it is rational to feel. Either way, however, our emotions are viewed as a mere epiphenomenon to the life of reason. Questions about what I *ought* to feel when, for example, I have betrayed a friend are completely submerged in, or displaced by, questions about what it would be most useful to feel.

A different, but closely related, view is used by Mellor in order to show how relief that a headache is over can be accommodated within a tenseless view of time. Mellor suggests that the relation between a painful experience and the relief which is felt when it stops is a purely causal one: of the same form as the relation between too much alcohol and hangovers. My relief is not relief *about* anything. So the claim that there is no fact—that is, the pain being over—for it to be about does not discredit this response. On Mellor's argument, then, our current responses are not saved from the demand for revision by being shown to be, despite appearances to the contrary, rationally justifiable. They are saved by being placed completely outside the domain of reason. Questions about what we *ought* to feel can no more arise with a response of relief than they can arise with the throbbing sensation in my head (Mellor 1981: 48–52).

While such views of the emotions are not uncommon within philosophy they do, of course, conflict violently with central strands of our normal thinking about the emotions. Certainly there is much in our emotional life which appears to be quite immune to the force of reason; and which, perhaps, most of us accept as best thought of as a condition to be treated, where possible, in much the way that a physical pain is to be treated. But there is also much in our emotional life which we think of quite differently. My resentment, gratitude, or fear may be judged appropriate or inappropriate in view of my circumstances; and the idea that what I felt was appropriate is to be sharply distinguished from the idea that it was the feeling which will have the greatest practical benefits. My resentment is misplaced if what I took to be malice was in fact a misfired attempt to be kind. Resentment is also misplaced if, as a result of my mood at the time, it is quite out of proportion to the severity of the offence. Relief is misplaced if there is worse to come, and sorrow over a minor loss may slip into self-indulgence. None of these judgements is undermined by the observation that what I felt was the most *useful* thing for me to have felt at the time.

III

The extent to which a philosopher sees the tenseless view as in conflict with our normal thought turns, perhaps, on the centrality which is given to the life of *feeling*. Spinoza, for example, whose picture of our ethical lives gives a fundamental place to the emotions, takes it to be clear that a full assimilation of a tenseless view has radical implications. I have suggested that this is correct: that the tenseless view inevitably carries a commitment to fundamental changes in our emotional lives; or, at least, a revision of fundamental significance in our understanding of the place of feeling in our lives. If the depth of this conflict has not been generally appreciated in recent discussions of these issues this may be a reflection of the rather peripheral place which the emotions have occupied in recent philosophy, and of our readiness to think of the emotions as non-rational forces, related to their objects in purely causal terms. I want now to return to *arguments for* the tenseless view, suggesting that two familiar recent defences of the view may depend for their apparent force on a failure to give the life of emotion its proper place in the discussion.

Mellor writes:

The sole function of tensed facts is to make tensed sentences and judgments true or false. But that job is already done by the tenseless facts that fix the truth-value of all tensed sentence and judgment tokens . . . Their tenseless truth conditions leave tensed facts no scope for determining their truth-values. But these facts by definition determine their truth-values. So in reality there are no such facts. (Mellor 1981: 102)

The argument bears a superficial similarity to certain arguments which purport to show that so-called 'secondary qualities' are not genuine features of reality:

We can explain our experience of objects as coloured without assuming that colour, as we experience it, is an objective property of objects. But the objective property of colour is by definition that which causes our experience of objects as coloured. (Alternatively: but there is no other reason for supposing that colour is an objective property of objects.) So in reality there are no such properties. (Or: so we have no reason to believe that there are such properties.)

The similarity is, however, only superficial. The latter argument appeals to an observed feature of our experience—our experience of objects as coloured —which, it is suggested, the idea of objective colour properties is intended to explain. By contrast, Mellor is not, I take it, suggesting that the following is an observed feature of our experience: tensed sentences and judgements are true or false. For what feature of our experience of the world could be thought to show that that is so? The point, I take it, is more of this form:

The idea that tensed sentences and judgements are true or false is one which is taken as simply given in all of our normal thought about past and future. That idea does not, however, commit us to the view that there are tensed facts. But this is the only role which tensed facts might be thought to play in our thinking. So in reality there are no such facts.

To this, however, it can be replied:

This is *not* the only role which tensed facts might be thought to play in our thinking. The idea that what has happened in the past, and what will happen in the future, can provide us with reasons to do certain things, and feel certain emotions, is also one which is taken as simply given in all of our normal thought about past and future.

Further, a central aspect of that is the idea that *what* an event of a certain kind gives us reason to do or feel depends crucially on whether the event lies in the past, present, or future: a past pain, death, or shameful deed has, we take it, a quite different bearing on our current thinking from that of a future one. But, if the argument of Section II was correct, that idea *cannot* be accommodated within a tenseless view.[6]

I do not present this argument as a defence of the claim that there are 'tensed facts'. For one thing, I feel considerable unclarity about just what that claim amounts to. For another, it is open to an advocate of the tenseless view to argue that the feature of our normal thinking to which I am appealing is a confusion. My point is just that Mellor's defence of the tenseless view contains a suspect premiss. Its plausibility turns on the assumption that the *fundamental* role which 'past and future facts' play in our lives is to be found in the idea that judgements in the past and future tenses can be true or false. Indeed, it appears to turn on the more radical assumption that this is the *only* role played by the idea of 'past and future facts'. It should come as no surprise that an argument based on that assumption leads to the conclusion that the different kinds of significance which we attach to past, present, and future are suspect.

Nagel's argument may raise slightly more awkward issues:

what can be asserted significantly of the present can be asserted significantly of other times also, and can be true or false of those times in the same sense in which it is true or false of the present. The present is just a time among others, and confers no special status on the circumstances which occupy it. . . . In sum: the sense of an assertion about what is the case at a certain time does not change with the tense of the assertion; the tense merely indicates a relation between the time of utterance and the time of what is being talked about. (Nagel 1970: 61)

Reflection on the fact that we can assert of other times the same thing as we assert of the present reveals, then, that tenses do not mark a deep division in reality. We can, however, ask: why should we not transpose this argument? We will then reason:

The *use* of sentences in the past and future tenses differs significantly from the use of their present-tense equivalents, in that, for example, the emotions for which the words 'I will suffer severe pain', 'I am suffering severe pain', and 'I did suffer severe pain' can intelligibly be given as a reason are radically different from each other. This shows that what can be asserted of the present—'I am in pain'—*cannot* be asserted in the same sense, cannot be true or false in the same sense, of other times.

---

[6] It will, I think, make no difference to my argument if one prefers to formulate the matter in the following way. The only role of tensed facts is to make tensed judgements, properly understood, true or false. But one criterion of whether tensed judgements *are* being properly understood is whether the states of affairs which are purported to make them true would serve to render the relevant actions and feelings appropriate. And the argument of Section II shows that the tenseless theory does not satisfy this criterion.

Nagel's defence of the claim that the tense of an utterance 'merely' indicates whether the event spoken of is earlier than, at the same time as, or later than the utterance rests on the position which his argument gives to one (supposed) aspect of the place which tense has in our thought, to the complete exclusion of other aspects: to the exclusion, indeed, of just those aspects—namely, the ways in which the tense of an utterance bears on how it may feature in the statement of reasons for feelings—which cast serious doubt on that 'merely'.

Of course, this leaves us with the question of whether the conclusion of the transposed version of Nagel's argument can possibly be right. Must we not accept that there is some content to the idea that what can be asserted of the present can be asserted, in the same sense, of past and future? Well, perhaps we must. But the views which we have been considering presuppose that that content can only be understood according to a certain model: a model which appeals to the idea of a common core to sentences in the past, present, or future tenses which have a common subject matter. That common core might be grasped independently of any mastery of tenses; or at least, independently of any mastery of the past and future tenses. Thus, we grasp what it is for rain to fall—or perhaps, as in Spinoza, what it is for rain to be falling now—independently of a grasp of what it is for it to be *going* to rain, or to *have* rained. And my mastery of the future-tense statement 'It will rain,' or the past-tense statement 'It has rained,' can be viewed as the result of bringing together this grasp with a general, independent, grasp of the future or past tense.

If the considerations presented in this chapter are correct then we need to explore the possibility of other models for the idea that the *same* thing can be asserted to *have* happened, to *be* happening, and to be *going to* happen. A possible starting-point for such an exploration would be the following remark by Wittgenstein:

'But surely "I believed" must tell of just the same thing in the past as "I believe" in the present!'—Surely $\sqrt{-1}$ must mean just the same in relation to -1, as $\sqrt{1}$ means in relation to 1! This means nothing at all. (Wittgenstein 1958: 190)

We might build on this starting-point, exploring the relations between the sense of differently tensed judgements, by asking: is it clear, as Nagel assumes, that the reason that I have to hope that I behaved with restraint at the party *derives from* reasons that I had at the time to avoid idiotic behaviour?

IV

In conclusion, I should stress that nothing that I have said is intended as a contribution to a refutation of the tenseless theory. I have suggested that the tenseless theory is committed to a demand for significant revisions in our thought about the emotions; and that certain defences of the theory depend on an assumption which we do not need to accept: the assumption that we can give a complete account of the sense of tensed language while making no mention of the contribution which the tense of an assertion makes to how it may feature in the offering of reasons for action or feeling. This, however, leaves open the possibility of a different form of defence of the tenseless view: one in which the call for a revision in our emotional lives is not presented as a *consequence* of a purely 'metaphysical' argument, but rather as something for which an independently compelling case can be made out and so which can *underpin* the demand for metaphysical revision. I suspect that some of Spinoza's thinking on these issues would be more accurately represented in that light.

REFERENCES

Ayer, A. J. (1954), 'Statements about the Past', in *Philosophical Essays* (London: Macmillan): 167–90.
Le Poidevin, R. (1991), *Change, Cause and Contradiction* (London: Macmillan).
MacBeath, M. (1983), 'Mellor's Emeritus Headache', *Ratio*, 25: 81–8.
Mellor, D. H. (1981), *Real Time* (Cambridge: Cambridge University Press).
Nagel, T. (1970), *The Possibility of Altruism* (Princeton: Princeton University Press).
Prior, A. N. (1959), 'Thank Goodness That's Over', *Philosophy*, 34: 12–17.
Seddon, K. (1987), *Time: A Philosophical Treatment* (London: Croom Helm).
Spinoza, Benedict de (1955), *Ethics*, trans. R. H. M. Elwes (New York: Dover).
Wittgenstein, L. (1958), *Philosophical Investigations*, ed. G. E. M. Anscombe and G. H. von Wright, trans. G. E. M. Anscombe (Oxford: Basil Blackwell).

5

Real Times and Possible Worlds

HEATHER DYKE

There are ways in which David Lewis's theory of genuine modal realism[1] is analogous to a tenseless token-reflexive account of time. At the semantic level Lewis offers an indexical account of actuality which is formally analogous to an indexical account of temporal terms such as 'now' and 'present'. At the ontological level there appears to be an analogy between Lewis's account of all possible worlds being equally real, and the doctrine of the tenseless theory that all times are equally real. My concern in this essay is to examine this apparent analogy between these two theories. How strong is it? How closely does it bind the two theories together? Is it strong enough to commit a tenseless theorist, by parity of reasoning, to genuine modal realism? My attempt to answer these questions will involve a close examination of each theory, an analysis of some attempts to undermine the analogy, and a consideration of some attempts to reinforce it. Ultimately I hope to prove that, as a proponent of the new tenseless token-reflexive theory of time, I am not forced to countenance Lewis's plurality of worlds.

1. The new tenseless token-reflexive theory of time

Originally espoused by Mellor (1981),[2] this is the theory that, although tense is ineliminable from thought and language, nevertheless it does not constitute part of temporal reality. So the theory does not aim to provide an analytic reduction of all tensed sentences to tenseless sentences. It recognizes that this is not possible. Instead, it claims to provide an ontological reduction of

I would like to thank Robin Le Poidevin, John Divers, and an anonymous referee for extremely helpful comments on this topic.

[1] In this essay, where I refer to modal realism, it is *genuine* modal realism I intend to discuss, except where I indicate explicitly that I am discussing *ersatz* modal realism.

[2] A version of the new tenseless theory of time was put forward by Smart (1980), but this was a *date* version rather than a *token-reflexive* version of this theory. For more on this distinction see my (1996).

tense to tenseless temporal relations.[3] This is achieved by giving the truth-conditions of tensed sentences in entirely tenseless terms. The token-reflexivity of these truth-conditions explains how different tokens of the same tensed sentence-type can have different truth-values. A token of 'The sun is rising' uttered before dawn is false, as is a token of the same type uttered at noon. It is only tokens of this type uttered simultaneously with the rising of the sun that are true. The metaphysical implications of this truth-condition project are that, although tense is a fundamental aspect of language and thought, it does not exist in reality. All that is needed to account for the objective truth or falsity of tensed sentences is tenseless temporal relations between events. Thus, the distinction between past, present, and future constitutes no part of reality whatsoever, and all times are ontologically on a par.[4]

2. *David Lewis's theory of genuine modal realism*

As I see it, there are two essential components of David Lewis's theory of modality. First of all, there is the indexical theory of actuality, which is given its fullest expression in Lewis (1970). According to Lewis, 'actual' is an indexical term and should be subject to an analysis of the same kind as that appropriate for other, more familiar indexicals such as 'here', 'now', and 'I'. The appropriate analysis for these indexicals is that their reference is wholly determined by, and varies according to, particular features of the context in which they are uttered. The reference of the indexical expression 'here' varies according to the place at which it is uttered or inscribed, so a token of 'here' refers to the place at which it is produced. Tokens of 'I' refer to the person who produces them, so they vary according to a different feature of the context of utterance: the utterer. Similarly, 'now' refers to the time at which it is uttered or inscribed, so non-simultaneous tokens of 'now' refer to different times. What then, is the appropriate feature of the context of utterance according to which the reference of the supposed indexical term 'actual' systematically varies? Lewis writes:

I suggest that 'actual' and its cognates should be analyzed as *indexical* terms: terms whose reference varies, depending on relevant features of the context of utterance. The relevant feature of context, for the term 'actual', is the world at which a given utterance occurs. According to the indexical analysis I propose, 'actual' (in its primary sense) refers at any world *w* to the world *w*. (Lewis 1970: 184–5)

---

[3] For further discussion on the different kinds of reduction that can be achieved by giving truth-conditional analyses of tensed sentences, and for further argument in favour of the essentially ontological nature of this particular kind of truth-conditional analysis, see my (1996).

[4] For a fuller discussion of this theory, see Le Poidevin's essay in this volume.

If, as Lewis claims, 'actual' is an indexical expression, such that its reference varies from world to world, then, he maintains, we can explain why scepticism about our own actuality is absurd:

How do we know that we are not the unactualized possible inhabitants of some unactualized possible world? . . . The indexical analysis of actuality explains how we know it: in the same way that I know that I am me, that *this* time is the present, or that I am here. All such sentences as 'This is the actual world,' 'I am actual,' 'I actually exist,' and the like are true on any possible occasion of utterance in any possible world. That is why skepticism about our own actuality is absurd. (Lewis 1970: 186)

Another consequence of the indexical theory of actuality is the rejection of the thesis that actuality is an absolute, non-relational quality. The indexical theory of actuality denies that what is actual is, in some sense, ontologically superior to what is merely possible. If the indexical theory of actuality is true, it is not the case that one world, the actual world, is more real than all the other merely possible worlds. For Lewis, this is a benefit of the indexical theory of actuality as it complements the other component of his position: modal realism.

Modal realism is 'the thesis that the world we are part of is but one of a plurality of worlds, and that we who inhabit this world are only a few out of all the inhabitants of all the worlds' (Lewis 1986: p. vii). Each world in this plurality is as real and concrete as every other world, although they are causally and spatiotemporally isolated from each other. Other worlds differ from the world we happen to inhabit, not in kind, but only in what goes on at them. The world we inhabit might have been different from the way it is, an intuition that we frequently express by the use of modal sentences, for example 'The pyramids might have been built by aliens.' According to Lewis, this sentence, if true, is made true by there being a world in which the pyramids (or a counterpart thereof) *were* built by aliens.

Genuine modal realism has one immediately obvious disadvantage: it generates an ontological inventory that is uneconomical in the extreme. There are many who would not countenance such a vast multitude of existents. Lewis is aware of this opinion, admitting that 'modal realism *does* disagree, to an extreme extent, with firm common sense opinion about what there is' (Lewis 1986: 133). However, he argues that its ontological disvalue is outweighed by its pragmatic value: there are 'many ways in which systematic philosophy goes more easily if we may presuppose modal realism in our analyses' (Lewis 1986: p. vii). What, then, are these pragmatic reasons for adopting modal realism; in what ways does it assist the progress of philosophy?

One reason in favour of genuine modal realism is that it disposes of the question 'Why should anything exist?' A world is a 'maximal mereological sum of spatiotemporally interrelated things' (Lewis 1986: 73). On that analysis there is no world where there is nothing at all, so it would seem to be necessary that there is something. Lewis does not claim to provide a genuine explanation of *why* there is something rather than nothing. It is merely a feature of his account that it turns out to be necessary that something exist, thus disposing of the question by making it irrelevant.

A more powerful reason in favour of genuine modal realism is that it affords remarkable clarity to analyses of modal expressions, and to modal reasoning.

> The standards of validity for modal reasoning have long been unclear; they become clear only when we provide a semantic analysis of modal logic by reference to possible worlds and to possible things therein.... Modal reasoning can be replaced by nonmodal, ordinary reasoning about possible things.... Once we have a nonmodal argument we have clear standards of validity; and once we have nonmodal translations of the premises, we can understand them well enough to judge whether they are credible. (Lewis 1970: 175)

The problem that Lewis identifies here, and purports to solve, is that there are many ways in which modal reasoning is abstruse if we take modal expressions such as 'possibly' and 'necessarily' to be primitive and irreducible sentential operators. For one thing, modal contexts are intensional such that co-referential terms cannot be substituted within them while the sentence as a whole retains the same meaning, or even the same truth-value. However, a theory that transforms opaque modal locutions into transparent ordinary locutions about modal things (possible worlds and possibilia) has the advantage of 'extensionalizing' modal reference and inference. Under this approach to modal reasoning the modality is contained within modal objects, so that we can refer to them and quantify over them in just the same way as we do for ordinary objects. According to this analysis, a sentence of the form 'Actually *p*' is true if and only if *p* is true in the world of utterance; 'Possibly *p*' is true if and only if *p* is true in at least one world;[5] 'Necessarily *p*' is true if and only if *p* is true in all possible worlds; and 'Not possibly *p*' is true if and only if *p* is true in no possible world. This approach yields an analogy with temporal expressions which quantify over times in a similar way. 'Actual' is thus analogous to 'present'; 'possible' is analogous to any tensed expression which locates an event at some time other than the present; 'necessarily' is analogous to 'always' or 'at all times'; and 'not possibly' or 'impossibly' is analogous to 'never' or 'at no time'.

---

[5] For Lewis, 'Possibly *p*' is true if and only if *p* is true in at least one world *accessible from* the world of utterance. The accessibility relation, within Lewis's theory, is one whose purpose is to define different modalities by delineating subsets of all the possible worlds. I will ignore the accessibility relation in this essay, as I am concerned with possible worlds *per se* and not with particular subsets of them. This relation does not contribute significantly to the discussion concerning whether there is an analogy between time and modality.

We now have before us a clear picture of Lewis's modal realism and his indexical theory of actuality, as well as of the tenseless token-reflexive theory of time. These brief excursions through the main tenets of each theory have already revealed some of the ways in which these two theories might be thought to be analogous. In the next section I will spell out this alleged analogy in more detail.

3. Tenseless time and 'modeless' modality

The tenseless token-reflexive theory of time offers a token-reflexive, or index-ical, analysis of tensed language. According to this analysis present-tense sentence-tokens are true if and only if they occur simultaneously with what-ever they are about. The truth of such tokens requires only that they stand in the appropriate tenseless relation to the event they refer to. It does not, as some tensed theorists would maintain, require that the events or times referred to possess any non-relational property of presentness. Nor does it require that the relevant events or times occupy some uniquely elevated ontological status, such that the present is more real than the past or the future. Every event and moment, according to this analysis, is present relative to itself. Thus, the token-reflexive analysis of tensed language is consistent with a tenseless ontology whereby all times are equally real, and none is uniquely distinguished from all the others by some characteristic of presentness. It does not entail that time is tensed in any way. A tensed theorist who wished to adopt the token-reflexive analysis of tensed language would be obliged to provide independent arguments for the conclusion that time is tensed, together with arguments for the conclusion that a tensed ontology is consistent with the token-reflexive analysis of tensed language. The two components of the tenseless token-reflexive theory of time are thus consistent with one another and, indeed, complement each other.

The two components of genuine modal realism appear to be consistent with one another, and to complement each other in precisely the same way. If one espouses an indexical analysis of the term 'actual', such that each world is actual relative to itself, is one thereby committed to the view that all worlds are equally real? If the indexical theory of actuality is true, then it is not the case that actuality is a metaphysically privileged status that one world, our world, attains. Analogously, if the status of presentness consists simply in the token-reflexivity of present-tense terms and the temporal indexical 'now', then it is not the case that presentness is a metaphysically privileged status that one moment, this moment, attains. I maintain that there is a close connection between an indexical analysis of tensed language and a tenseless

ontology. This connection consists in the fact that, in the absence of arguments
to support a tensed ontology, together with arguments to defeat McTaggart's
claim that a tensed ontology is inherently paradoxical, the only temporal ontology that can consistently be maintained in conjunction with a token-reflexive
analysis of tensed language is a tenseless one. The token-reflexive analysis
of tensed language seems naturally to imply a tenseless ontology according
to which all times are equally real. If there is an equally close connection
between an indexical analysis of actuality and an ontology of genuine possible
worlds, then a question naturally arises concerning whether a token-reflexive
analysis of actuality naturally implies an ontology of worlds according to which
all worlds are equally real. If the formal analogy is robust then an argument
can be generated to the effect that a tenseless theorist with respect to time
is committed, by parity of reasoning, to genuine modal realism. My main
objective in this chapter is to undermine any suggestion that a proponent of
the tenseless theory of time is committed by analogy to genuine modal realism. As a first step towards that objective I will examine some attempts by
other philosophers to undermine the apparent analogy.

4. An examination of some attempts to undermine the analogy

There have been some attempts to uncouple these particular conceptions of
time and modality from each other. Various strategies have been adopted in
order to achieve this end. E. J. Lowe, for example, argues that there is a
significant disanalogy between times and possible worlds that is sufficient to
undermine any suggestion that the two theories may be analogous in any way.
He describes this disanalogy as

An important dissimilarity between temporal relations and relations between possible
worlds, consisting in the fact that the former put instants of time into a unique linear
order not paralleled by any analogous linear ordering of possible worlds. In short,
time constitutes a *dimension* in a way that possibility does not. (Lowe 1986: 195)

Essentially the argument is that there is a lack of ordering in the domain of
possible worlds, and that this constitutes an intrinsic difference between worlds
and times, such that they cannot be assimilated to one another, or treated in
a similar metaphysical fashion. That there is a disanalogy here at all has been
denied by Yourgrau (1986), and of course Lewis,[6] who both argue that the
relation of similarity between worlds is sufficient to provide them with an
ordering, even if we are unable, in principle, to determine precisely what that
ordering is.

[6] See, for example, Lewis (1973).

It seems to me that this objection, and the response to it, both miss the mark. Whether or not possible worlds constitute a dimension seems to be irrelevant to the question of whether our reference to our own world, and reference to other worlds by their inhabitants, with the use of the expression 'actual' succeeds via an indexical mechanism.[7] If we take the terms 'now' and 'I' as being, fairly uncontroversially, indexical expressions, we can see clearly that whether the domain of objects referred to by these expressions is ordered or not is irrelevant to their indexical nature, and is irrelevant to the question of whether every object in the domain is on the same ontological level.

The reference of the expression 'now' varies according to when it is uttered, and the reference of the expression 'I' varies according to who utters it. The domain of reference for 'now' is constituted by times which, as Lowe pointed out, are ordered by the temporal relations in which they stand to each other ('earlier than' and 'later than'), in such a way as to constitute a dimension. The domain of reference for 'I' is constituted by speakers of a common language, and although spatiotemporally related to each other, they are not ordered by any 'personal' relation analogous to the temporal relations, and do not constitute a dimension. None of this prevents 'I' from functioning successfully as an indexical, nor does it provide any grounds for denying that all speakers of a language exist on the same ontological level. So, whether the domain of reference of an indexical expression is well ordered (as in the case of times), or not (as in the case of speakers of a language), does not appear to have any consequences for whether that expression qualifies as an indexical, or for whether the objects in that domain of reference are on the same ontological level. This objection has indeed focused on a dissimilarity between times and possible worlds, but I maintain it is an irrelevant dissimilarity. That is, it has no consequences for the existence or otherwise of a semantic or metaphysical analogy between genuine modal realism and any theory of temporal reality.

Graeme Forbes adopts a rather different strategy in his attempt to undermine the suggestion of an analogy between time and modality. He argues that our experience of time differs significantly from our experience of modality. He concludes from this that, metaphysically, times are not analogous to possible worlds. Consequently there is no threat of there being analogous theories of time and modality. We may infer from this that if the tenseless theory of time and modal realism appear to be analogous, this is no more than an appearance because there are significant underlying differences between the categories of time and modality. The difference that Forbes chooses

---

[7] Although this is not the point that concerned Lowe in the article I quoted from.

to describe is that 'we move through time . . . but do not move through the other possible worlds in which we exist' (Forbes 1983: 127). Is this difference enough to unhitch tenseless time from modal realism?

For the tenseless theory of time the claim that we move through time is a mere figure of speech used to convey the truth that we have different temporal parts at different times. Thus, if the difference identified by Forbes is decisive enough to undermine any supposed analogy between times and possible worlds, we ought to cash it out in terms consistent with the tenseless theory of time and modal realism. In these terms Forbes's disanalogy reduces to the claim that we have different temporal parts at different times, but we do not have different 'modal' parts at different possible worlds. Restating the disanalogy in these terms results in its dissolution. According to modal realism we have many counterparts that exist in many different possible worlds. Analogously, according to the tenseless theory of time, we have many different temporal parts that exist at many different times.

However, there are some tenseless theorists who deny that individuals have temporal parts, and altogether reject temporal part theory. It may be that Forbes's disanalogy is more effective in divorcing tenseless time from modal realism if understood in terms consistent with the tenseless theory of time that rejects temporal part theory. According to such a theory, the notion of movement through time is an allusion for an individual's standing in different temporal relations to different events and moments of time. For us to recognize an event, e, as being first future, then present, and then past, is just for us to judge at some time, t, that e is later than t, and at some later time t_1 that e is simultaneous with t_1, and at some still later time t_2 that e is earlier than t_2. Within such a theory Forbes's disanalogy reduces to the claim that we stand in different temporal relations to different times, but we do not stand in analogous relations to different possible worlds. Once more the alleged disanalogy dissolves. Tenseless time and modal realism are indeed analogous in this respect. Just as we stand in different temporal relations to different times, other possible worlds are related to the world we inhabit in virtue of being more or less similar to our world. Employing the disanalogy that Forbes fixes on to attempt to release the tenseless theory of time from its association with modal realism is unsuccessful.

Martin Davies's strategy looks far more promising in the bid to uncouple tenseless time from modal realism. He argues that the indexical theory of actuality is false. 'Actual' is not an indexical expression, according to Davies, so there is no analogy between it and 'now' or 'present'. If Davies's argument is successful, the allegedly analogous reasoning does not even get past the first premiss. Davies writes:

If a speaker uses 'now' . . . in an utterance, then it is not sufficient for understanding of that utterance that an audience should merely know the meaning of the sentence uttered. He must be able to identify the time . . . of the context of utterance; only then will he know what has been said . . . But one can hardly hold that understanding of an utterance involving 'actually' requires identification of the actual world from amongst the set W of possible worlds. (Davies 1983: 132)

The force of this disanalogy lies in the claim that indexicals typically change their reference from one context of utterance to another, and that to understand how the reference changes is to understand the indexical expression. The argument continues that while this is true for indexicals, it is not true for the term 'actually' because its reference does not change in our experience; it always refers, for us, to the world we inhabit. To understand this term then, we do not have to identify the context in which it is uttered, and so it is not a genuine indexical. A similar objection is made by Peter van Inwagen (1980), who argues that treating 'actually' as an indexical is a gratuitous and illegitimate extension of the notion of indexicality, for much the same reasons as those offered by Davies. My response is equally applicable to both arguments.

Based on the notion of what it is for an expression to be an indexical, this objection begs the question against the indexical theory of actuality. It is true that all utterances in this world of 'This is the actual world' are true in this world, but this is in accordance with Lewis's position. It does not follow from this that utterances in other worlds of the same sentence-type are not true in those worlds in virtue of the term 'actual' referring indexically to those worlds. This objection only has any force if one has already decided that 'actually' and its cognates do not function as indexicals. It can only be claimed that this is a gratuitous extension of the notion of indexicality if one provides solid grounds for that claim. If the only reason offered is merely the assumption that the claim is true, then the objection has no force whatsoever.

I believe that the arguments I have discussed in this section are representative of objections to the existence of an analogy between time and modality. They singularly fail to undermine any such analogy. That being so, the analogy not only appears at first sight to be compelling, it is also able to withstand the objections that have so far come its way. I turn in the next section to scrutinize one particular formulation of the analogy put forward by M. J. Cresswell. D. H. Mellor defends McTaggart's argument for the unreality of tense, and adopts a tenseless theory of time as an alternative to McTaggart's own conclusion that time is unreal. Cresswell argues that if Mellor's argument for the tenseless theory is successful, and there are many who believe that it is, then so too is an equally powerful argument for David Lewis's theory of genuine modal realism.

5. A modal version of McTaggart's paradox

According to McTaggart tense is unreal because the supposition of its existence yields a contradiction. If tense is real then the following two propositions must be true:

(1) Past, present and future are incompatible determinations. Every event must be one or the other, but no event can be more than one.
(2) But every event has them all. If M is past, it has been present and future. If it is future, it will be present and past. If it is present, it has been future and will be past. (McTaggart 1927; repr. in Le Poidevin and MacBeath 1993: 32)

These two propositions clearly contradict one another, so McTaggart concludes that tense is unreal. Since he believes that tense is also essential to the reality of time itself, he concludes that if tense is unreal, so too is time. Mellor draws a different conclusion from this contradiction. Since tense is unreal, but tense is not essential for the existence of time, time exists and is tense-less.[8] M. J. Cresswell constructs a modal analogue of McTaggart's paradox with the aim of proving that, just as tense is unreal, primitive modality is unreal, and consequently modal reality must consist in a plurality of Lewisian worlds. The modal analogue goes as follows:

Many M-positions are incompatible with each other. An event which is merely possible for example cannot also be actual. Being merely possible and being actual are mutually incompatible properties of things and events. But because they are contingencies everything has to have them all. Everything occupies every M-position from merely possible to actual. But nothing can really have incompatible properties, so nothing in reality has modal properties. M-positions are a myth. (Cresswell 1990: 165–6)

First of all, what are we to understand by an 'M-position'? Cresswell intends it to be the modal analogue of an A-series position, which is a position in time identified as past, present, or future. Let us suppose, then, that an M-position is a position in 'modal space', identified as actual, merely possible, impossible, necessary, and so on. However, if this is the correct interpretation of an M-position, then the argument is quite disanalogous to McTaggart's temporal argument. There are many M-positions that are quite compatible with each other. An event may be both actual and necessary, actual and contingent, or possible and necessary. So, it would seem that, for many of these M-positions, there is really no difficulty in saying that events can occupy more than one of them. McTaggart, on the other hand, is quite right to say that every A-position is mutually incompatible with every other A-position.

[8] For a fuller discussion of McTaggart's paradox, see Le Poidevin's essay in this volume.

However, it is possible that Cresswell intends to talk only of actuality and mere possibility as representing M-positions. Since these two M-positions are incompatible, the argument goes through, up to this point at least, although it remains puzzling just what Cresswell means when he says, 'Everything occupies every M-position from merely possible to actual.' What M-positions are there between actual and merely possible?

The next move in McTaggart's argument, after establishing that every A-position is incompatible with every other A-position, is to note that every event occupies every A-position. Because the tenses of events are forever changing, every event has to have them all. The modal counterpart of this move in Cresswell's argument is to say that because M-positions are 'contingencies', every event has to have them all. Cresswell appears to imply that there is an analogy between the continual change of tense that events undergo, and the contingency of modal properties. If there is an analogy here at all it is a weak one, certainly too weak to support the inference to the claim that everything occupies every M-position. Furthermore, is it really the case that M-positions are occupied contingently? Modalities are generally taken to be ascribed necessarily and not contingently; for instance, it is necessarily necessary that 2+2 = 4. Is Cresswell suggesting that we dispense with this practice so that mathematical truths are merely contingently necessary?

In seeking to construct a modal analogue of McTaggart's paradox, Cresswell faltered when it came to invoking a modal analogue of the continual change of tense that events undergo. He appealed to the 'contingency' of modal properties, but it clearly does not validate the analogous modal inference. Thus, this modal analogue of McTaggart's paradox fails to force one into the position either of rejecting modality as incoherent (as McTaggart rejects tense as incoherent) or of adopting modal realism (as Mellor adopts tenseless time). It fails, I believe, because there is no clear modal analogue of the change of tense that events and times appear to undergo. Modal realism may be analogous to tenseless time in many respects, but it is not the case, as Cresswell claims, that a modal analogue of McTaggart's paradox commits us to modal realism. The analogy may be powerful, but it is not that powerful. However, it may still be powerful enough to compel a tenseless theorist to adopt modal realism or admit inconsistency. In order to avoid facing that choice I turn now to examine Lewis's theory a little more closely.

6. Token-reflexivity and real worlds

Lewis's conception of modal reality consists in his genuine modal realism together with the indexical theory of actuality. Aware that the major disadvantage of his position is its extreme ontological requirements, Lewis invites us to peruse its advantages, which are largely pragmatic, and to consider whether the ontological price is worth paying. Many of its pragmatic benefits derive from its association with the indexical theory of actuality. But what precisely is the relation between the two components of his theory? Does the indexical theory of actuality entail modal realism, or does modal realism entail the indexical theory of actuality? I maintain that neither entailment relation obtains.

To see that the indexical theory of actuality does not commit us to a plurality of worlds, consider the following example. According to the indexical theory of actuality, a token of the sentence 'It is actually the case that the sky is blue' is true if and only if the sky is blue in the world in which that token is produced. So far there is no commitment to the existence of any world except the world in which the token occurs. What if things had been different? What if the sky had been khaki? In that case the expression 'the actual world' would have denoted a different world if the indexical theory of actuality were true. However, this does not commit me to the existence of a world where the sky, as seen from a counterpart of the planet earth, is a uniform shade of khaki. One can adopt the indexical theory of actuality even if the expression 'actual' only applies to one world.

To illuminate this restriction, it might be useful to consider the indexicality of the pronoun 'I' as understood by a solipsist who believes that, as a matter of fact, no other minds exist apart from her own. Sally the solipsist will be quite happy to adopt an indexical analysis of 'I' such that it refers, on each occasion of use, to the one who utters it, even though there is, as a matter of fact, only one speaker that this indexical expression ever successfully refers to: herself. The fact that Sally believes herself to be the only person who genuinely exists will not cause her to redefine the meaning of 'I' such that it is an alternative name for herself. She already has a perfectly good name. No, the indexical analysis of 'I' is the most useful and explanatory account of that expression, despite the effective limitation in its genuine reference.

Thus, one can hold the indexical theory of actuality without thereby being committed to genuine modal realism. It is noteworthy that one can also adopt genuine modal realism without thereby being committed to the indexical theory of actuality. One can hold that there is a plurality of genuine worlds, except that one of them differs from all the others by possessing the absolute, non-relational quality of actuality. Such a theory is attributed by Armstrong to Leibniz, although he notes that 'the textual warrant for this is dubious' (Armstrong 1989: 3). According to a theory of this kind,

Over and above the actual world there are an indefinite multiplicity of merely possible worlds. They constitute all the ways the world could have been ... It is natural to develop this view by saying that it involves *two levels of being*. The actual world has the superior sort of being: actuality. The merely possible worlds have some sort of being, but they lack actuality. (Armstrong 1989: 3–4)

Arguably, a proponent of such a theory faces even more difficulties than Lewis. Not only does this theory have the ontological drawbacks of Lewisian modal realism, but it must also provide an account of that in virtue of which *this* world is actual, and not some other world. Indeed, on this view, our world might not be actual and we may be deluded in thinking that it is. It must also account for the fact that it seems to involve two levels of being: actual existence and merely possible existence.

The indexical theory of actuality and genuine modal realism can be prised apart from one another. Opponents of genuine modal realism can adopt the indexical theory of actuality, with all its attendant pragmatic benefits. One might object that, as a theory for illuminating modal expressions of ordinary language, the indexical theory of actuality is effectively rendered useless without the notion of possible worlds. Arguably it *would* be rendered ineffectual without the notion of possible worlds, but we can employ the notion without admitting their concrete existence. We simply need to find some ontologically benign substitute for Lewis's possible worlds. There are a number of alternative theories of modality which employ the notion of possible worlds, but which stop short of recognizing their concrete existence. For example, those who espouse what Lewis calls *ersatz* modal realism take possible worlds to be abstract entities that represent ways this world might have been. They have been construed as, among other things, maximal consistent sets of sentences (Stalnaker 1976), as true stories (Adams 1974), and as idealized representational pictures. An alternative account of the nature of possible worlds is given by Armstrong (1989), who suggests that we should see possible worlds as useful fictions, very much like the notion of the ideal gas, the average family, or the economic man. We can usefully employ these notions, distinguish true from false statements about them, yet there is no need or reason to postulate their very existence. I do not mean to take up any particular one of these theories of modality, I merely wish to point out that there exist more palatable alternatives to Lewisian modal realism.

The fact that the two components of his theory can be divorced from one another in this way will not deter Lewis from retaining modal realism. He has presented a substantial number of arguments against the various alternative accounts of possible worlds which suggests he has no intention of giving up his own conception of modal realism in favour of such a theory. In fact, it is more likely to be the case that his primary commitment is to modal realism, from which he takes an indexical account of actuality to follow:

This makes actuality a relative matter: every world is *actual at* itself, and thereby all worlds are on a par ... The 'actual at' relation between worlds is simply identity ... Given my acceptance of the plurality of worlds, the relativity is unavoidable. I have no tenable alternative. (Lewis 1986: 93)

Lewis will still adopt modal realism, even though it is not entailed by the indexical theory of actuality, believing it to have, on balance, greater pragmatic value than any competing alternative account of modality, or any alternative account of the nature of a possible world. However, I believe that genuine modal realism faces a serious problem.

Lewis claims to offer us a completely reductive analysis of modality. He maintains that it is a positive advantage of his account that it leaves no unanalysed modal primitives awaiting explanation. Bearing this in mind, let us examine how he accounts for the modal notion of impossibility. Nathan Salmon argues that 'whatever grounds there may be for believing that there really are possible worlds yield the same, or related reasons for believing that there are impossible worlds' (Salmon 1984: 116). Margery Bedford Naylor (1986) and Takashi Yagisawa (1988) have argued that Lewis's own arguments for possible worlds, if successful, work *mutatis mutandis* for impossible worlds. For example, Yagisawa argues:

There are other ways of the world than the way the world actually is. Call them 'possible worlds'. That, we recall, was Lewis' argument. There are other ways of the world than the way the world could be. Call them 'impossible worlds'. That is the extended argument [which] has exactly the same form as Lewis' original argument. (Yagisawa 1988: 183)

In other words, the reasoning which Lewis invokes to support the existence of possible worlds also provides support for the existence of impossible worlds. On what grounds then does Lewis deny the existence of impossible worlds? Since Lewis refuses to include impossible worlds among his plurality of worlds, Lycan has argued (1991*a*) that his concept of a world invokes a modally primitive notion.

Lycan's reasoning is that if an argument for the existence of a possible world yields an equally good argument for the existence of an impossible world, then his rejection of impossible worlds and retention of possible worlds depends on our antecedent understanding of the notion of 'possible'. The concept of a possible world is itself dependent on an unanalysed modality. This is a serious problem for Lewis because he claims to offer us a completely reductive analysis of modality if only we will accept his plurality of worlds.

He urges that this is what makes his theory superior to any ersatz modal realism, because the latter theories must retain some modal notion as primitive. If, however, it turns out that his theory too retains a modal notion as primitive, then, ontologically, we ought to reject his theory.

Richard B. Miller defends Lewis by offering the following as a definition of 'world' that supposedly appeals to no modal primitive:

(1) Individuals are worldmates if they are spatiotemporally related.
(2) A world is a mereological sum of worldmates. (Miller 1989: 477)

Lycan responds:

Since nothing but the impossibility of round square cupolas in the first place keeps a round square cupola from being spatiotemporally related to another, perhaps less exotic object, Miller's Lewis would not be able to rule out *worlds* containing round square cupolas. (Lycan 1991*b*: 212)

Miller is not convinced:

'Round square cupola' purports to be a description of a possible object, but it is not. It is a contradictory description and so describes nothing. Hence it describes no individual. Hence it describes no individual having spatiotemporal relations. Hence it describes no worldmate. So there is no need to amend the definition of world to limit worlds to possible worldmates. (Miller 1993: 159)

I submit that Miller's response does not succeed in overturning Lycan's objection. He claims that 'round square cupola' does not describe a possible object because it is a contradictory description. What does Miller mean by 'contradictory' in this context? Perhaps he means that it is not possible for any individual to possess both of these properties at the same time. If so, then he has failed to find a non-modal reason for excluding impossible objects, and hence impossible worlds, from the modal landscape. Miller anticipates this response and defends his use of the term 'contradictory':

Genuine modal realism provides the resources to define 'contradictory' without the primitive modal operator. We require only quantification over worlds and define 'contradictory' as false in every world. (Miller 1993: 160)

At this point I contend that Miller has led the argument full circle without ultimately defeating the objection. 'Contradictory' means false in every world simply because Miller and Lewis reject the existence of impossible worlds. If there were any impossible worlds, then 'contradictory' would have to be defined as 'true in some impossible world'. Miller and Lewis must reject the existence of impossible worlds because their stipulation that worlds are concrete, physical things prevents there being any impossible ones. However, this rejection of impossible worlds is based on the prior distinction between possible and impossible—a modal distinction. I conclude that there is an inherent circularity in Lewis's genuine modal realism because he cannot eliminate modally primitive notions from the account, while claiming to provide a completely reductive analysis of the modalities. This difficulty, taken together with the extreme ontological disvalue of genuine modal realism, renders it a highly problematic account of modality.

The circularity inherent in Lewis's theory is a subtle one. His position is not overtly circular as would be the case if his reductive analyses of modal sentences themselves contained irreducibly modal notions. For example, the modal sentence 'Possibly there are flying elephants' is true according to Lewis's analysis if and only if there is a world at which elephants fly. As we have seen, a world is a mereological sum of spatiotemporally related worldmates. Consequently, an analysis of modal sentences such as this in terms of worlds understood in this way does not contain any irreducibly modal notions. The inherent circularity emerges not in the individual analyses of modal sentences, nor in the notion of what a world *qua* concrete entity consists in, but in the range of worlds supposed by Lewis to exist. The notion of a world in itself does not contain any irreducibly modal notions, but in designating those worlds that exist, Lewis is forced to rely on the modal distinction between possible and impossible. Thus, the circularity in Lewis's position exists at the level of the specification of which worlds exist. It is in delineating the range of worlds over which his existential quantifier operates that Lewis implicitly relies on irreducibly modal notions.

The difficulties I have identified with Lewis's modal realism are at least twofold. First of all, the pragmatic value of Lewis's theory as a whole derives largely from the indexical theory of actuality together with some notion of possible worlds over which one can quantify. However, as it turns out, one can adopt these highly illuminating aspects of Lewis's position without having to take on board his genuine modal realism. The indexical theory of actuality is what one might call a promiscuous theory, able to attach itself to any coherent account of the nature of a possible world. Given any such coherent account, we can also quantify over possible worlds, thus retaining these two important benefits, but dispensing with the drawback of a plurality of genuine worlds.

Lewis himself does not agree with that conclusion. He believes that the alternative accounts of possible worlds, on balance, run into more problems than does his own account of them. He asks us to grit our teeth and embrace an ontology of genuine possible worlds. It is worth it, he tells us, for only by doing so can we achieve a completely reductive analysis of modality. However, the Miller–Lycan problem reveals that it is only apparently completely reductive. There are primitive modalities hidden away among those possible worlds. On the whole then, I choose to reject genuine modal realism. However, it remains to be seen whether I can consistently do so while retaining a tenseless and token-reflexive account of time.

7. *Token-reflexivity and real times*

Earlier in this chapter it became clear that there are reasons to suppose that a tenseless token-reflexive theory of time is closely analogous to Lewis's genuine modal realism. In the previous section I identified some serious problems for the latter theory. My task in this section is to consider whether the problems that infected Lewis's theory have temporal analogues that do analogous damage to the tenseless token-reflexive theory of time.

The first difficulty for Lewis was that the indexical theory of actuality and quantification over worlds can both be adopted independently of a genuine ontology of possible worlds. These two beneficial aspects of his theory can be prised apart from its problematic aspect. Can an indexical account of 'now' and 'present', together with quantification over times, be similarly prised apart from a genuine tenseless ontology whereby all times are equally real?

A token-reflexive analysis of tensed sentences appeals to the existence of events, times, and the temporal relations of precedence, subsequence, and simultaneity. For the tenseless theory, these temporal relations are constitutive of time. Thus, the fact that the token-reflexive analysis of tensed language requires their existence gives us reason to suppose that the semantic and the ontological aspects of this theory complement each other. However, there are at least two tensed theories that purport to adopt a token-reflexive analysis of tensed language in conjunction with a tensed ontology. These are the theories of Quentin Smith and E. J. Lowe. The existence of these theories suggests that, perhaps, a token-reflexive analysis of tensed language is not as closely tied to a tenseless ontology, according to which all times are equally real, as tenseless theorists would like to think.

Quentin Smith's theory of temporal reality incorporates a token-reflexive analysis of tensed language into an ultimately tensed theory of time.[9] He argues that tensed sentence-tokens have token-reflexive rules of usage, but that this is insufficient to ensure the truth of a tenseless theory of time. Indeed, he argues, the hypothesis that tensed sentences express non-relational tensed propositions is not only consistent with these tenseless rules of usage, but also explains why tensed sentences obey them. For example, on this account,

[9] For a fuller account of Quentin Smith's theory of time see Le Poidevin's essay in this volume.

a token *T* of the tensed sentence 'The storm is approaching' is true if and only if both *T* and the storm's approach are present. That is, they both possess the irreducible, non-token-reflexive property of presentness (Smith 1993: 98). If these events are both present, it follows that the token-reflexive truth-conditions of *T* also obtain. That is, *T* is simultaneous with the storm's approach.

Now, what makes Smith's theory tensed is his inclusion in it of tensed, non-token-reflexive properties, such as pastness, presentness, and futurity. However, there is nothing in the token-reflexive analysis of tensed language itself which entails the existence of such properties. Independent reasons need to be given to invoke their existence. In my view, Quentin Smith's position does not threaten to divorce the token-reflexive analysis from a tenseless ontology. Rather, Smith is introducing an additional category of temporal entity into his ontology, a move which leaves the onus firmly with him to provide justification for it. Furthermore, his account must overcome the problems inherent in adopting a tensed ontology. That is, Smith must defeat McTaggart's argument that real tense is inherently self-contradictory.

The theory espoused by E. J. Lowe purports to accommodate the token-reflexivity of tensed sentences. Lowe offers token-reflexive truth-conditions for tensed sentences that relate utterances to the events they are about according to the temporal relations of 'earlier than', 'later than', and 'simultaneous with'. However, what makes Lowe's theory tensed is that the times at which these events occur, and between which these relations obtain, are A-series times rather than B-series times. As he remarks, 'I don't believe that times can ultimately be specified in purely tenseless terms' (1993: 173). As I noted earlier, the ontological requirements of the token-reflexive analysis of tensed language are that times, events, and temporal relations exist. Lowe's theory, it seems, fulfils those requirements, leaving nothing out and smuggling nothing additional into the ontological inventory. The token-reflexive analysis requires the existence of times, but it is silent as to the requisite nature of those times. It seems that there is nothing to prevent Lowe from adopting the token-reflexive analysis in conjunction with an ontology of A-series times. Just as the ersatz modal realists are free to adopt the indexical theory of actuality in conjunction with their own account of the nature of possible worlds, so Lowe is free to adopt the token-reflexive analysis of tensed language together with his own account of the nature of times.

Lowe's theory seems so similar to the tenseless token-reflexive theory of time that one is tempted to ask why he advocates the reality of the A-series rather than the B-series. What is the point of maintaining that the A-series is real? The point is, of course, to support the claim that there is a real, non-relational distinction between the past, present, and future. To attest the reality of this distinction is not merely to recognize a semantic distinction, but to hold that there is a real, ontological disparity between what can truly be described as past, present, and future. The belief that there is such an ontological distinction is usually expressed as the belief that there is an onto-logical asymmetry between the past, the present, and the future. Indeed, Lowe admits to recognizing some such distinction when he says:

Unlike the definite description 'the past', the definite description 'the future' is non-denoting, by virtue of the fact that there are (barring determinism) many possible futures, no one of which is ontologically privileged, but only one (actual) past. (Lowe 1993: 173)

It seems to me that there is nothing in the token-reflexive analysis of tensed language that would support the existence of such an ontological asymmetry. Indeed, it is arguable that the token-reflexive theory requires the reality of all times. If a future-tense sentence has truth-conditions that relate the pres-ent utterance to the future event, then the fulfilment of those truth-conditions arguably requires the existence of that future event to stand in that temporal relation. I submit that Lowe must supplement his theory in the following way. He must provide independent support for the ontological asymmetry between past, present, and future, together with arguments to the effect that the token-reflexive analysis is consistent with such an asymmetry. Furthermore, he too must overcome arguments such as McTaggart's for the conclusion that real tense is inherently self-contradictory. Hence, I contend, Lowe's position is similar to that of Smith. Each of them has yet to prove that the token-reflexive analysis is consistent with a tensed ontology. However, its ontological require-ments show that it is most naturally associated with a tenseless ontology.

In the preceding discussions I do not claim to have defeated either Smith's theory or Lowe's. What I hope to have shown is that, unlike Lewis's oppon-ents, these two theorists have not managed to drive a wedge between the semantic aspect of the tenseless token-reflexive theory and its ontological conclusions. What they have both done is to attempt to supplement the onto-logical commitments of the token-reflexive analysis with additional tensed entities. I have argued that there is insufficient reason to suppose that tem-poral ontology is lacking without these additional tensed items, and that the onus remains with them to prove that it is.

I intimated above that Lowe's strategy is rather similar to that of the ersatz modal realists who argue that they can employ possible worlds discourse while offering their own account of the nature of possible worlds. Lowe claims to employ a token-reflexive analysis of tensed language while offering his own account of the nature of times. However, there is a significant difference between these two strategies. Ersatz modal realists, as well as other philosophers who employ the notion of possible worlds without admitting their concrete existence, claim that possible worlds are not worlds in the sense that *this* world is. They take worlds to be entities of a wholly different kind, for example maximal consistent sets of sentences or useful fictions. I do not think Lowe would maintain that past and future times are entities of a wholly different kind from the present time. However, there is a theory of the nature of temporal reality that advocates just such a position. It is worth taking a look at this position because it has an obvious modal counterpart.

The position with respect to time known as temporal solipsism holds that only the present exists. As Prior put it: 'The present simply *is* the real considered in relation to two particular species of unreality, namely the past and the future' (Prior 1970: 245). The parallel with the modal position known as actualism should be clear. According to actualism only one world exists, the actual world, other possible worlds being a mere species of unreality. Temporal solipsism has received its fullest treatment to date in Le Poidevin (1991). As a theory of time it poses a potential threat to the tenseless token-reflexive theory of time in a way that the theories of Smith and Lowe failed to do so. It threatens to drive a wedge between the token-reflexive analysis of tensed language and a genuine tenseless ontology. We saw earlier that it is quite consistent to adopt an indexical analysis of actuality, such that utterances of 'actual' refer indexically to the world in which they occur, while holding that, as a matter of fact, only one world exists. Similarly, it is quite consistent to adopt an indexical analysis of 'I' even if, as a solipsist, one holds that there is only one existent to which 'I' ever genuinely refers. Temporal solipsism is able to exploit this feature of indexicality, adopting an indexical analysis of 'now' while holding that only the present moment genuinely exists. If this theory is genuinely workable, then it represents a radical schism between a token-reflexive analysis of tensed language and a tenseless ontology.

Temporal solipsism, however, is not a genuine alternative to the tenseless token-reflexive theory of time. In his (1991) Le Poidevin issues several decisive blows against temporal solipsism, rendering it, in my view, untenable as a theory of time. In what follows I will examine one of Le Poidevin's arguments against temporal solipsism, which illustrates how this theory actually differs substantially from its apparent modal counterpart, actualism.

The temporal solipsist is faced with the task of providing a reductionist account of instants. Since he maintains that the past and the future are unreal, he must either 'regard all talk of past and future instants as contentless, which is absurd' (Le Poidevin 1991: 54), or else he must 'reconstrue such talk' (ibid.). Le Poidevin examines various alternative entities which the temporal solipsist may choose to identify with past and future instants, but concludes that

The only remotely plausible reductionist strategy open to him is to identify instants with *propositions*: an instant is the conjunction of propositions which would ordinarily be said to be true at that time. (Ibid.)

Such a propositional theory of instants is quite analogous to that particular brand of ersatz modal realism which construes possible worlds as maximal consistent sets of sentences. However, the propositional theory of instants faces a serious problem. The temporal solipsist takes past and future times to be identical with conjunctions of propositions and, according to Le Poidevin, must do so. It follows that he cannot, without circularity, hold that times are part of the content of those propositions. Thus, two tokens of the same tensed sentence-type uttered at different times must express the same proposition. This is because all that differentiates them is the time at which they are uttered, but it is not open to the temporal solipsist to invoke this difference in order to distinguish them from each other. As Le Poidevin remarks:

On the solipsist's picture, then, different tokens of the same tensed type, *even when uttered at different times*, express the same proposition. The consequence is that the very same tensed token (e.g. this very inscription of 'Socrates is sitting') can be true at one time and false at another. This makes nonsense of tensed assertion. (Le Poidevin 1991: 55)

There is no analogous problem for theories of ersatz modal realism which take other possible worlds to be, for example, maximal consistent sets of sentences. The very same sentence-token can be located at different worlds, under this picture of modal reality, and possess different truth-values at some of those worlds. The ersatzer does not have to invoke 'occurrence at a world' to distinguish between different tokens of the same sentence-type. This is because there is nothing incoherent in supposing that different tokens of the same type express the same proposition in different worlds and yet differ in truth-value in those worlds. Their individual truth-values depend on what is the case in the world in which they are uttered, which, of course, differs from world to world. Consequently, it is perfectly coherent to maintain that any two such tokens express the same proposition and yet differ from each other in truth-value. All this, in my view, provides more reason to let go of the idea of an analogy between time and modality. The 'commonsense' modal position of actualism has a far-fetched and ultimately untenable temporal analogue.

To conclude this section I will briefly consider whether the tenseless token-reflexive theory of time suffers from a difficulty analogous to the circularity problem that infected Lewis's genuine modal realism. Lewis claims to offer a completely reductive analysis of modality, although the Miller–Lycan discussion revealed that his theory in fact retains the modal distinction between possible and impossible as primitive. Lewis is forced to deny the existence of impossible worlds. There is no temporal analogue of this problem. Within Lewis's theory there are modal reasons for rejecting the existence of impossible worlds. There are no temporal reasons for rejecting the existence of non-existent times. Such times might have existed. There is no temporal principle which rules out non-existent times, as there is a modal principle which Lewis employs to rule out non-existent worlds. Thus, the tenseless theory of time does not suffer from the same kind of circularity as Lewis's modal realism because, whereas Lewis smuggles irreducibly modal notions into his analysis of modality, there is nothing irreducibly tensed in the ontology of the tenseless theory of time.

To put the point more simply, Lewis claims to offer a completely reductive analysis of modal notions in terms of possible worlds discourse. The new tenseless theory of time, by contrast, does not claim to offer a completely reductive analysis of tense in terms of a tenseless language. The claim of the new tenseless theory of time consists merely in an ontological reduction of tense to tenseless relations, but it retains tense as an irreducible feature of language and thought. Thus, the new tenseless theory of time cannot be charged with claiming to offer a completely reductive analysis of tense while retaining irreducibly tensed notions in the account, since it does not claim to provide a completely reductive analysis of tense.

An alternative interpretation of a temporal analogue of Lewis's circularity problem might be as follows. Lewis's rejection of impossible worlds could be construed as a rejection of worlds which are inaccessible from the world of utterance, or worlds that are modally disconnected from the world of utterance. Furthermore, the arguments of Section 4 showed that Lewis is forced to deny the existence of such worlds, but he has no real, non-modal justification for rejecting them. The temporal analogue of the problem construed in this way would arise if the tenseless theory of time is forced to reject the existence of times which are temporally disconnected from the time of utterance, while having no justification for rejecting such times. However, the possibility of the existence of temporal series which are temporally disconnected from each other is not an incoherent one. Indeed, there is a sense in which the existence of this possibility supports the tenseless theory of time over the tensed theory.

If there exists more than one temporal series, each temporally disconnected from the others, then it is conceptually coherent to conceive of the events and times within those series being temporally related to each other by the tenseless relations of precedence, subsequence, and simultaneity. However, it is not easy to make sense of the distinction between past, present, and future applying to those series. This is because, arguably, one needs to be temporally located *within* a temporal series in order coherently to apply the concepts of past, present, and future to the events and times of that series. Thus, one cannot coherently apply the concepts of past, present, and future to a temporal series that is temporally disconnected from the temporal series in which one is located. This supports the conclusion that the concepts of past, present, and future are not intrinsic to a temporal series, but merely dependent on one's being temporally located within it. By contrast, the temporal relations of 'earlier than', 'later than', and 'simultaneous with' are, arguably, intrinsic to a temporal series, since their application to such a series is not dependent on one's being located within it.[10]

To sum up the conclusions of this section, I argued that tensed theories such as those offered by Lowe and Smith do not threaten to divorce the token-reflexive analysis of tensed language from a tenseless ontology in the same way that alternative accounts of modality can divorce possible worlds discourse from an ontology of genuine modal realism. Temporal solipsism offered, at first sight, a more palpable threat to the tenseless theory of time, since it is capable of adopting the token-reflexive analysis of tensed language in conjunction with an ontology in which only the present exists. However, I argued, with Le Poidevin, that temporal solipsism is ultimately untenable as a theory of time. Finally I considered whether the circularity problem that infected Lewis's account of modality had a temporal analogue that might undermine the tenseless theory of time. I argued that, under one possible account of such an analogy, there is no problem generated for the tenseless theory of time. Under another possible account the temporal analogue generated, far from being incoherent, actually supports the tenseless theory, while creating serious problems for the tensed theory.

8. Conclusion

It seems at first sight that there is a compelling analogy between genuine modal realism and the tenseless token-reflexive theory of time. A consideration of some earlier attempts to undermine any such analogy showed that they failed, thus giving us more reason to suppose that the analogy is robust. For a tenseless theorist who did not wish to be compelled, by parity of reasoning, to adopt genuine modal realism, things were looking grim. I then examined an attempt to construct a modal analogue of McTaggart's paradox, and

---

[10] In his (1996) Le Poidevin argues that the possibility of disunified time is harder to reconcile with the tensed theory than with the tenseless theory. Currie (1992) provides further support for this conclusion by arguing that the events in a fictional time-series, and thus in one temporally disconnected from our own, are ordered by the temporal relations, but do not have tenses.

concluded that it failed. Hence, one route towards the adoption of a tense-less ontology did not yield an analogous modal path towards Lewis's genuine ontology of possible worlds. A detailed analysis of Lewis's position revealed that his indexical theory of actuality does not entail, nor is it entailed by, his possible worlds ontology. These two elements of his theory can be prised apart. For those who endorse the indexicality of actuality, but reject a plurality of worlds, this is good news, but Lewis himself continues to endorse the plurality of worlds. It then became clear that genuine modal realism fails in its quest to provide a completely reductive analysis of modality, because its treatment of the notion of impossibility reveals that primitive modal notions are retained within the account. All this is reason enough for me to reject Lewis's genuine modal realism. It remained for me to consider whether these lines of thought spawned analogous reasons to reject the tenseless theory of time.

I examined the theories of Smith and Lowe which purport to combine a token-reflexive analysis of tensed language with a tensed ontology. Neither theory managed to divorce token-reflexivity from a tenseless ontology. Instead, they sought to introduce additional tensed items into their temporal ontology. In each case more argument is needed to accept the existence of such items. Temporal solipsism initially appeared to pose more of a threat to the tenseless ontology, with its ability to combine a token-reflexive analysis with the ontological position that only the present exists. The arguments of Le Poidevin proved this position to be untenable. Finally I examined whether tenseless time suffers from an analogous circularity problem to that which infects Lewis's modal realism, and concluded that it does not.

In conclusion, I admit that the analogy between these two theories appeared cogent, but I have argued that this was only an appearance of cogency. Proponents of the tenseless token-reflexive theory of time do not find themselves unwittingly committed by analogy to the concrete existence of a plurality of worlds.

REFERENCES

Adams, R. M. (1974), 'Theories of Actuality', *Noûs*, 8: 211–31.

Armstrong, D. M. (1989), *A Combinatorial Theory of Possibility* (Cambridge: Cambridge University Press).

Cresswell, M. J. (1990), 'Modality and Mellor's McTaggart', *Studia Logica*, 49: 163–70.

Currie, Gregory (1992), 'McTaggart at the Movies', *Philosophy*, 67: 343–55.

Davies, Martin (1983), 'Actuality and Context Dependence II', *Analysis*, 43: 128–33.

Dyke, Heather (1996), 'A Philosophical Investigation into Time and Tense' (Ph.D. dissertation, Department of Philosophy, University of Leeds).

Forbes, Graeme (1983), 'Actuality and Context Dependence I', *Analysis*, 43: 123–8.

Le Poidevin, Robin (1991), *Change, Cause and Contradiction: A Defence of the Tenseless Theory of Time* (London: Macmillan).

—— (1996), 'Time, Tense and Topology', *Philosophical Quarterly*, 46: 467–81.

—— and MacBeath, Murray (1993) (eds.), *The Philosophy of Time* (Oxford: Oxford University Press).

Lewis, David (1970), 'Anselm and Actuality', *Noûs*, 4: 175–88.

—— (1973), *Counterfactuals* (Oxford: Basil Blackwell).

—— (1983), *Philosophical Papers*, vol. i (New York: Oxford University Press).

—— (1986), *On the Plurality of Worlds* (Oxford: Basil Blackwell).

Lowe, E. J. (1986), 'On a Supposed Temporal/Modal Parallel', *Analysis*, 46: 195–7.

—— (1993), 'Comment on Le Poidevin', *Mind*, 102: 171–3.

Lycan, William G. (1991*a*), 'Two—No, Three—Concepts of Possible Worlds', *Proceedings of the Aristotelian Society*, 95: 215–27.

—— (1991*b*), 'Pot Bites Kettle: A Reply to Miller', *Australasian Journal of Philosophy*, 69: 212–13.

McTaggart, J. M. E. (1927), 'The Unreality of Time', repr. in Le Poidevin and MacBeath (1993: 23–34).

Mellor, D. H. (1981), *Real Time* (Cambridge: Cambridge University Press).

Miller, Richard B. (1989), 'Dog Bites Man: A Defence of Modal Realism', *Australasian Journal of Philosophy*, 67: 476–8.

—— (1993), 'Genuine Modal Realism: Still the Only Non-circular Game in Town', *Australasian Journal of Philosophy*, 71: 159–60.

Naylor, Margery Bedford (1986), 'A Note on David Lewis' Realism about Possible Worlds', *Analysis*, 46: 28–9.

Prior, A. N. (1970), 'The Notion of the Present', *Studium Generale*, 23: 245–8.

Salmon, Nathan (1984), 'Impossible Worlds', *Analysis*, 44: 114–17.

Smart, J. J. C. (1980), 'Time and Becoming', in P. van Inwagen (ed.), *Time and Cause: Essays Presented to Richard Taylor* (Dordrecht: D. Reidel Publishing Company): 3–15.

Smith, Quentin (1993), *Language and Time* (New York: Oxford University Press).

Stalnaker, Robert C. (1976), 'Possible Worlds', *Noûs*, 10: 65–75.

van Inwagen, Peter (1980), 'Indexicality and Actuality', *Philosophical Review*, 89: 403–26.

Yagisawa, Takashi (1988), 'Beyond Possible Worlds', *Philosophical Studies*, 53: 175–204.

Yourgrau, Palle (1986), 'On Time and Actuality', *British Journal for the Philosophy of Science*, 37: 405–17.

6

Time as Spacetime

GRAHAM NERLICH

1. Introduction

The sore spot in the philosophy of time is, and always was, the idea that the present time is metaphysically remarkable whereas the present place is not. So the present time is, somehow, a burning topic, one which is tied, in the minds of many, to the vexed question whether or not time flows. There are two main but rather different ideas about all this: first, that it has something to do with language; second, that it is tied tightly to the way we experience the world. The first idea has led to a great deal of ingenious analysis of tensed language aiming either to show that, metaphysically, tense is deep or that it is shallow. The second theme has proved less approachable for, while the sense of something experientially primitive and striking about the present is strong, it proves difficult indeed to say what this highly intuitive feature is. I want to discuss both sorts of view.

Quentin Smith's *Language and Time* (1993) makes some new and power-ful claims about time and the present. Smith's presentism is the view that presentness is a logical subject of every proposition and a metaphysical sub-ject of every state of affairs. He argues, that is, that every proposition is tensed. Each true one corresponds to a state of affairs of which presentness is a part. The present is, therefore, a universal subject of both propositions and states of affairs—moreover, it is the only universal subject. He argues that numbers and propositions exist in time, so that mathematical and logical statements also refer to presentness. Thus the whole of reality is unified in presentness —including presently past and presently future things and states of affairs (§ 8.1). Clearly, this does not appeal to intuitions about temporal experience. Indeed, as Smith insists (1993: 135), it is counter-intuitive. I admire the boldness of ideas, the brilliance of invention, and the density of argument in his book.

I have just described part II, which is written as semantics plus some syntax together with an austere but strong metaphysics. The way is paved for it in part I by an exhaustive examination of attempts either to translate tense out of ordinary language or to provide tenseless semantics for tensed discourse. That is, he attacks the work of B-series thinkers. B-theorists aim to show that tense is a superficial aspect of discourse, and thus that all temporal facts fit into a picture of time employing only the relations earlier, later, and simultaneous. Presentness would be demoted to a mere indexical, erased from our list of properties and deprived of a privileged status in time. Smith argues that all these detenser strategies fail. He supports a real A-series of past, present, and future. I dodge the critical arguments in part I but tackle the main semantic and metaphysical aspects of the theory. I will argue that Smith is mistaken about them.

If the relativity theories are correct in their standard interpretations, then time is an aspect of spacetime and 'present' properly functions in temporal language indexically, just as it does in spatial language. I argue that the Special Theory of Relativity differs from Smith's view of time more profoundly than he seems to think. Finally I suggest part of what, in experience, makes the idea of a real present so attractive.

2. *Basic semantic apparatus*

I shall begin by sketching a methodological background, quite uncritically, since it is Smith's own. My strategy will be to show that Smith's methods commit him to propositions and states of affairs in which the property of presentness plays no part. I raise no question at all about the correctness of the methods.

First, I mention a main strategy, even though it is not my target. F. P. Ramsey thought it obvious that 'Socrates is wise' is logically identical with 'Wisdom inheres in Socrates.' Smith *argues* the point that every sentence is logically identical with one of the form 'Presentness inheres in such and such', where '( ) inheres in such and such' expresses a second-order property of the property *presentness*. He uses Ramsey's identity to jockey presentness into subject position. Logical identity for sentence-tokens is as follows (1993: 137): (S1) and (S2) are logically identical iff they refer in the same way to the same items and ascribe to these items the same n-adic properties. I shall not be concerned with that part of Smith's case. However, it explains why he gives predicates a referring role. That shapes much of his syntax and semantics, inside which I will work.

Consider the 'paradox of the list': if the function of every syntactic part of a sentence is to refer to something, then the sentence can amount only to a *list* of items—it fails to identify a proposition, either true or false. So every proper sentence contains a syntactic part which does not refer. Searle's familiar reply to this conundrum is that predicates do not refer.

Smith's solution is different (1993: 139–40). Names refer to things and predicates to properties; there is no need to distinguish these types of entity for the beginning purpose of semantics. However, it cannot be the sole function of every syntactic part of a sentence to refer. In particular the syntactic part which we call the copula has two distinct functions (1993: 142). The important one, for us, is that it *conveys* something (as against referring to it). What it conveys is that all the items in the sentence which are referred to by its speaker stand in a propositional relation. But it also refers to presentness.

A distinction goes along with this. Consider sentence-tokens. If a syntactic part of a sentence refers to something, then that thing is a *part* of the proposition expressed. Parts of propositions are also constituents of them, but there is always one (non-referring) *constituent* which is not a part (1993: 141). It is the propositional relation, which orders the parts to one another so that a truth or falsehood results. Without such a constituent we have mere lists.

Smith's main example (we will need no other) has three aspects: the tensed sentence-token (an A-sentence) 'John is running,' the proposition which it expresses, and the state of affairs to which it corresponds if true. Thus states of affairs, here, are what others call facts. Note that propositions, states of affairs, and correspondence are so conceived that the propositional relation, which is a constituent of P, is *replaced* by exemplification, the tie of property to thing, in the corresponding state of affairs. Notice, too, that propositions are not Fregean: it is John and running that stand in the proposition, not a concept of John nor a concept of running. For example, John in the proposition is *identical* with John in the state of affairs. The proposition has John and running as its parts and the propositional relation, which orders the parts into a proposition, as a constituent.

Consider Smith's account (§ 5.6) of states of affairs. 'A state of affairs is whatever corresponds to a true proposition' (1993: 151). 'The complex relation of correspondence is analyzable in terms of three simpler components, *identity, replacement*, and *instantiation* such that these latter connections obtain between constituents of the proposition (or, derivatively, of the sentence-token) and constituents of the state of affairs' (ibid.). We have learnt earlier that the pure copula 'is' conveys that 'the items introduced into the proposition by other parts of the sentence-token *stand in a certain relation, namely the propositional relation*' (1993: 142). And on p. 139, that 'the most important element in the definition of the propositional relation involves a reference to truth values . . . more exactly stated by saying that if several items are related by this relation, then these items as so related *originarily (non-derivatively)* possess a truth value'. (All italics in these quotes are Smith's.) These directions strongly suggest and, I shall argue, strongly commit Smith to the kind of structure illustrated in Fig. 1. S is a sentence, P a proposition, and SoA

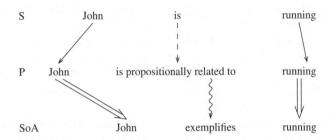

FIG. I

a state of affairs. The arrows between S and P represent reference, conveyance, and reference, respectively; those between P and SoA represent identity, replacement, and identity, respectively. The role of instantiation in correspondence is briefly explained in note 1 below.

3. *Presentness purged*

My objection to Smith is that his account of the truth of A-sentences (and of their correspondence to states of affairs) commits him to basic structures in which truth and correspondence falsify presentism. A basic structure contains both a proposition and a state of affairs which exclude presentness as a logical or a metaphysical subject (as is shown in Fig. 1). Smith's full account of A-sentences explicitly builds on this basic structure and fully incorporates it. So every fully explicated correspondence of an A-sentence with a state of affairs incorporates a tenseless sub-proposition and a tenseless sub-state of affairs.

Smith seems not to foresee this—neither that this basic structure is the obvious way for a reader to follow him nor that his account of correspondence fully commits him to it. He does not see that he tells a significantly different story about the truth of A-sentences, a story which goes unmotivated and unexplained in ways crucial for the success of presentism. He introduces the structure I speak of but misdescribes it. So I will look closely at his own discussion of the truth of A-sentences to illustrate this.

Smith assumes (1993: 152) that some token of the A-sentence

    John is running

expresses a proposition which is true. Note that, since it is present-tensed, it refers to presentness. Crucially, Smith defers this aspect of it. He begins with these items:

John   is propositionally related to   running   (P1)

and

John   exemplifies   running   (S1)

This shows 'the connections that obtain between these propositional constituents and constituents of the state of affairs . . . where (P1) includes constituents of the proposition and (S1) constituents of the state of affairs that corresponds to the proposition if the latter is true' (ibid.). We can see that these are related by the components of the relation of correspondence specified on p. 151: John is identical with John, running with running, and '( ) *exemplifies* ( ) *replaces* ( ) *is propositionally related to* ( ) in the sense that the former [*sic*— but 'the latter'?] *connects* John to running in the proposition and is thereby the order among these parts in the proposition, whereas this ordering function is taken over by ( ) *exemplifies* ( ) in the state of affairs' (1993: 152). Exemplification is the property tie in states of affairs. All this, together with the very suggestive '(P1)' and '(S1)' labels, leads one to conclude that the former is a proposition and the latter a state of affairs to which this proposition corresponds as its truth-maker. It seems to be just Fig. 1 all over again. (All italics in this paragraph are Smith's.)

I claim that this is fully warranted and inescapable. To be sure, (P1) is not the proposition expressed by the token of the A-sentence. (P1) does not have presentness as a part. Nor, and for the same reason, is (S1) the state of affairs to which that token (derivatively) corresponds. Still, that can hardly be a reason for denying the status I claim for (P1) and (S1). Yet, if that is correct, presentism is false. There is a proposition, (P1), and a state of affairs, (S1), which do not have presentness as subject.

However, we may draw—and emphasize—a very general conclusion: what goes for 'John is running' goes for all tensed propositions. Smith presents the example to illustrate the structure of all present-tensed propositions and, clearly enough from this, the structure of all propositions however they are tensed. Similarly for states of affairs. If so, then Smith's semantics commits us to a general core structure of propositions and states of affairs, and of correspondence between them, which strongly falsifies presentism, since the core case is the centre of every case. Basic propositions and basic states of affairs always exclude presentness as a logical or a metaphysical subject. Every correspondence relation between a true tensed proposition and a state of affairs is built on a core tenseless structure.

Smith might claim, on the basis of his part I, that there is no such proposition and no such state of affairs since there is no sentence available by which to identify them. But even if we grant those arguments, that is not enough. Reality, as Smith notes in another connection, overflows the use of language

(1993: 134). Whether or not there is some state of affairs in the world can hardly be settled by whether or not we have linguistic devices to pick it out.

Smith takes no steps to explain what disqualifies (P1) as a proposition and (S1) as a state of affairs. We are simply told that the former is a propositional *complex*, and that it only *quasi*-corresponds to (S1), since the latter is an event (1993: 152). Despite the fact that (P1) and (S1) are related by John's identity with John, running's with running, and exemplification's replacing the propositional relation (all of which is offered as defining correspondence), an exemplification of a property by something is called an *event* (1993: 152). No reason is given for saying so. We learn that (P1) and (S1) are 'not complete propositions and states of affairs but propositional complexes and events' (ibid.). Obviously, once again, they are not complete *presentist* entities, but what disqualifies them as a proposition and a state of affairs respectively remains obscure. Especially obscure since (S1) is just the sort of thing which figures, on p. 134, as a state of affairs when the term is introduced: 'If a sentence-token ascribes a property F to some existent E and the token is true, then there corresponds to the sentence-token a state of affairs that consists of the existent E and the property F as ordered by exemplification.'

Now turn to the question for which all this work aims to prepare us, the question of which proposition our specimen token-sentence, present-tensed as it is, might express and which state of affairs it might (derivatively) correspond to. Smith now picks out (P2), which has presentness as a part, and (S2), which also contains the property. They are pictured thus (1993: 153).

| [ ] | is propositionally related to | presentness | (P2) |
| [ ] | exemplifies | presentness | (S2) |

The brackets are to be filled by (P1) and (S1) respectively.[1]

Now (S2), one might think, is the state of affairs to which the token A-sentence corresponds. But if we follow Smith's earlier directives ('an event is an exemplification of a property by something' (1993: 152)), this is not a state of affairs but an event. He seems to intend us to follow them, since

---

[1] We need two propositional relations, a complication I have deleted here. In (S1) John is identical with John in (P1), but what fills the bracket in (P2) is not identical with what fills the bracket in (S2). According to Smith, the event (S1) *instantiates* the propositional complex (P1). Instantiation (see above) is the third component of correspondence. (See Smith's 5.4 for an extended discussion of propositional C-relations.) My revisionary story runs like this: what stands in the proposition (P2) is not the state of affairs (S1), but the proposition (P1) which corresponds with (S1). On neither story is the simple E-propositional relation appropriate, since the item in the state of affairs is not identical with the item in the proposition, but instantiates it. Rather, it instantiates if we allow the correspondence relation between propositions and states of affairs to be a case of instantiation. Thus (S1) instantiates or co-exemplifies both (P1) and presentness. In short, this omitted complication is no help to Smith against my objection.

(foot of 152) the state of affairs we are looking for is not (S2) but S, although Smith nowhere describes or illustrates it. Nor do we see the proposition (P, one presumes) which corresponds to S as its truth-maker. But, now, what *is* a state of affairs? We are never shown a clear distinction between them and events.[2]

Is (S1) an event? An event is something in the same category as a running, John's running, for instance, or a falling or a breaking. Now (S1) is not a running, nor even an exemplifying; so it is not an event, but congruous with a proposition, congruous so as to correspond, not merely quasi-correspond, with one. Of course, the true proposition expressed by our A-sentence-token guarantees that there is indeed an event, John's running. (P1) also entails that. But neither (S1) nor (S2) is this event. Each is a state of affairs. There must also be an event of John's running becoming present.[3] However, that does not entail either that (S1) or (S2) is an event any more than it entails that the unseen S is an event.

Here we might expect an argument for a strong claim made earlier for presentism: 'It is by virtue of presentness inhering in something that states of affairs are constituted and that there are states of affairs rather than nothing at all' (1993: 135). We need to see what virtue presentness has whereby its inherence in things alone can constitute states of affairs. Then we might grasp why running's inherence in John, his exemplifying running, falls short of being a state of affairs. There is no such argument. Nothing, here or elsewhere, seems even intended to establish 'that there are states of affairs only by there being presences' (1993: 136). On the contrary, it is, throughout, by virtue of something's exemplifying a property that they are constituted. We learn, for instance, that the items

John, the property of running, the property of presentness

are a heap of individuals and properties, not a proposition (1993: 139); and that the mere aggregate

the property of being the tallest man, the property of running, the property of presentness

[2] Nor between propositions and propositional complexes. However, there is some need for subtlety here, though no need for some of the distinctions which Smith draws. (No presentist-neutral need, that is.) There is need to distinguish *exemplifies*, *co-exemplifies*, and *co-satisfies*. We need the last if we say that propositional complexes are not exemplified—they are not properties. We do not need events and quasi-correspondence but only states of affairs and correspondence, when one proposition contains another and one state of affairs contains another.

[3] The event is that John's running becomes present or exemplifies presentness and thus undergoes a change in its properties analogous to a change in temperature of a thing. This is just the kind of consequence of presentism which seems so repugnant to good sense.

does not form a proposition, although 'These items form a proposition by virtue of standing in a propositional relation' (1993: 144). There must be a constituent of the proposition (not a part of it) which orders these parts one to another so that they stand in a propositional relation. That presentness is among the items as a part has no proposition-forming virtue. Nothing is said about another role which presentness might play in forming propositions or states of affairs.

I conclude that Smith's own principles of semantics and metaphysics entail that (P1) is a proposition, and that (S1) is a state of affairs. So there are propositions which lack presentness as a logical subject and states of affairs which lack it as a metaphysical subject. Clearly, for most token A-sentences which are true, there are related propositions and states of affairs of which presentness is not a part. Presentism is false if these arguments are correct.

4. Tenseless propositions

Now for Smith's contention, in chapter 6, that presentness is a logical subject of tenseless sentences.

Let us return to the pure or grammatical copula, which is one of two syntactic parts[4] of a standardly used tensed token-sentence, such as 'John is running' (1993: 142). More precisely, it is what is common to 'is', 'was', and 'will be' in the variably tensed versions of the sentence above (ibid.). It is not the property of being tensed, but exemplifies it. It conveys, in our example, that a propositional relation holds between John and running.

We learn (1993: 181–3) that it is thin states of affairs which sustain the thesis of presentism. Thick ones allow lots of universal subjects, such as self-identity and oneness. Thinness applies, directly, to parts of states of affairs. 'Thin John is John taken apart from all his properties' (1993: 182) and he, devoid of all properties, necessary or contingent, is what stands in the thin states of affairs which concern us. Thinness must also apply to parts of propositions, too. If not, John in the proposition will not be identical with thin John in the state of affairs it must correspond to. Thus John sheds his necessary but thick trappings and only his nucleus gets in (1993: 184). A nucleus is part of a thick referent and exemplifies every other part. Yet, for a reason which is never given, exemplification, in states of affairs, and the propositional relation, in propositions, are always thick constituents. They cannot, it seems, be purged (as parts must be) of their necessary companions, presentness,

---

[4] The discussion of tense on p. 187 suggests just one syntactical part, however.

present pastness, and so on.[5] Why they cannot remains unexplained. Smith argues that there is no such reading of *sentences*, by claiming that the copula lacks a tense only in the weak sense that it lacks a single tense (1993: 188).

The argument occurs on pp. 190–2. Smith proposes to identify the semantic function (not just the purely copulative function) of 'is' in

Socrates is wise (1)

with the complex 'was, is, or will be' in

Socrates was, is, or will be wise (2)

so as to produce logically identical sentences. There are two reasons given for concluding that they are logically identical. First, both are intersubstitutable *salva veritate* in belief contexts; second, they fulfil a condition of *syntactico-semantic correlation*

if and only if, *consistently with their known semantic properties*, there is some way of dividing S1 and S2 into syntactic parts, such that for each part P1, distinguished in S1 there is one and only one part P2 in S2 that has the same thin semantic content as P1. The parts . . . may be simple or complex. (1993: 191)

Syntactico-semantic correlation looks set to deliver conflicting results. A crucial question is begged if we presuppose, as in the definition, that *each* part of the sentence has semantic content, does refer to something. Yet one syntactic part, in every sentence, conveys a mere *constituent* of a proposition. So we must settle a main question about reference and thin content before we know to which syntactic parts the condition on content applies. Does it apply to 'is'?

Next, imagine someone who thinks that spacetime is a universal metaphysical subject and who thinks that (1) is logically identical with

Socrates is somewhere, sometime, wise (3)

This passes the (vague, I admit) belief-context test. Now correlate the complex syntactic part 'is somewhere, sometime' with 'is' in (1). How may we decide whether or not they have the same thin semantic content? Presumably, (2) and (3) cannot have the same thin semantic content. (1) cannot be logically identical with both of them. We must now turn to the independent merits of presentness and spacetime as universal metaphysical subjects in order to settle syntactico-semantic correlation.

Similar arguments work against Smith's claim that numbers and propositions exist in time. I have no space to pursue them.

---

[5] I assume, for brevity of argument, that these are necessary companions.

5. *True time is spacetime*

Consider a new theme: that time is rightly understood as a dimension of space-time. If the preceding arguments are correct, then presentism is a misunderstanding. I argue that the time of STR (Special Theory of Relativity) is not reducible to classical relations, but rather the reverse.[6] I begin by looking at something which Smith says about this.

The opening arguments of *7.2 Metaphysical Time and Special-Theory-of-Relativity-Time* (1993: 229–30) suggest, plausibly, that there is a true concept of time, embedded in some ideology,[7] so that other concepts of time, in other ideologies, are false or derivative. There is only one true time. Smith offers (1993: 230) three criteria for choosing it:[8]

1 Events in true time are related by primitive, irreducible relations.
2 Events in pseudo times are reducible to non-temporal relations or to true temporal relations.
3 Events in pseudo times which are related to each other by pseudo relations are also related by true temporal relations.

It is easy to see that, if true time is STR time, condition (2) will be met by reducing common-language temporal relations to STR ones. For the former will be reducible to frame-relative temporal relations, just as *large* and *small* are reducible to ( ) *is larger than* ( ) and ( ) *is smaller than* ( ). Obviously, STR time meets conditions (1) and (3).

Smith believes that STR relations are definable in terms of luminal, causal relations. I disagree. STR is a theory about spacetime and what it contains, not, first, about light and causal relations. Note that Smith's causal definitions are modal, concerning connectibility, possible signal passing, etc. On what do these possibilities and impossibilities depend? They are not mere logical or metaphysical ones: it is logically and metaphysically possible for distant events, simultaneous relative to a given frame of reference, to be linked by a signal. Simply pick another frame and conceive of a faster-than-light signal relative to it. Had the speed of light been infinite, even simultaneous events could be linked by light.

Clearly, these definitions presuppose some kind of physical modality: they depend on physical laws. But which law or laws? That depends on how one

---

[6] Many of the arguments in this section are developed more fully in Nerlich (1994: chs. 2–4).

[7] I hope that 'ideology' may serve as a term which is neutral between a full explicit theory, such as STR, and a view built into the very structure of language, as Smith thinks his account of time is.

[8] Smith speaks of metaphysical time, tying the form of the discussion to his explication of 'exists in time'. I do not mimic fully the structure of his discussion.

formulates the theory. Smith's discussion requires that a fundamental postulate of STR be that nothing outstrips light; that is how the theory is about luminal connections and connectibility. But this is not the only (nor the most interesting) suggestion. Another view sees STR as placing an overriding structural constraint on all laws, the constraint of Lorentz invariance. The crucial thesis of STR is a second-order one. (See Friedman 1983: ch. iv, §§ 5, 6.)

If the second-order constraint, Lorentz invariance, lies close to the core of STR, that makes it a fundamental as well as a novel theory, since it shapes every other. However, the constraint itself suggests a still deeper interpretation, one which sees space and time by themselves as imperfectly perceived aspects of a single, unified entity, spacetime. Lorentz invariance springs from the symmetries of this all-embracing entity. That is what Minkowski taught us. This interpretation of STR is dominant.

Two episodes in modern theoretical physics illustrate the ascendancy of the Invariance Principle over the Limit Principle (which states that the speed of light is the limit of all speeds). The first of these is the speculative development of the theory of tachyons, or faster-than-light signals; the second is the possibility of the massive photon, which would entail that light *in vacuo* does not occupy the null cones of spacetime. Each hypothesis contradicts the Limit Principle in different ways. The (speculative) physics of each hypothesis is developed within STR as based on the Limit Principle. (See Feinberg 1967, Recami and Mignani 1974, Barnes 1979, and Goldhaber and Nieto 1971.) Therefore light connections cannot be used to reduce the temporal concepts of STR. They do not define the meaning of STR nor define its concepts.

If STR commits us to spacetime then it commits us to an absolute temporal relation, *after*—and this relation defines time. Early in the history of STR, Robb (1914) saw how to capture the whole structure of spacetime by the use of a single relation, which he called *after*. It is a surprising fact that, given just the *after* or *elsewhen* connections of every point of Minkowski spacetime to every other, we may deduce the whole congruence structure of it. This basis is to be preferred to the use of frame-relative temporal relations.

Robb explained his *after* relation by speaking of the possibility of causal action from the earlier to the later event. As before, we should note the modal in this explanation and ask for its source. The causal possibility can be used only to *illustrate* the temporal order, not to *found* it. Temporal relations cannot be analysed by causal ones since causal relations presuppose them.

However, the literature standardly refers to Robbian analyses of spacetime structures, in both STR and GTR, as causal and to *after* as causality. But this, I have argued, is inaccurate and regrettable. (See Nerlich 1994: chs. 2 and 3.)

6. *Time and intuitions*

Nevertheless, there is much about time which STR does not tell us. It says that presentism (of whatever sort) is false but sheds no light on why it attracts us. It says nothing about the flow or the direction of time. It says just enough to show that time is not as we are naturally inclined to think it.

Something in experience deludes us about time, something obvious and pervasive—there obtrusively in every thought and perception, yet elusive to analysis. We take it that this something would be there whenever and wherever we experience anything. No doubt its foundations lie in subtle or complex aspects of the physical world. But, in terms just of how things seem to us, there must be a useful (if subtle) answer to the question why the delusion of presentism draws us so strongly. Here is part of an answer.

It is significant that we do not perceive things as extended in time in the way that we see them as extended in space. I can see things as distant from where I see them and see them as occupying volumes which bear no special relation to the volume I fill (the volume of the place from which I see them). But I do not see things as happening earlier than when I see them, even when I know they are earlier. If I see an occlusion of Ganymede by Jupiter, then I see now what I know occurred many minutes ago. But, inevitably and ineducably, I see it as occurring now. If I see an extended happening, then I see it as taking the same time to happen as it takes me to see it. Its extension in time (temporal 'volume') seems exactly the same as that of my seeing it. So fast- or slow-motion photography looks comic because we see the happening itself as absurdly slow (fast). There is no perspective effect in temporal perception, whereas in spatial perception we are quite used to seeing as huge, distant mountains which fill but a small part of our visual field. Thus we may be led to think, in the familiar paradox, that we do not really see astronomical events since we do not see them when they occur. But we have no inclination to think that we do not see distant mountains because they are not where we are, as we see them.

Thus we tend to take facts about the structure of our perception of time for facts about the ontic structure of present, past, and future.[9] We perceive only what happens while we perceive it; we remember what has happened; we merely guess (sometimes educatedly) at what will happen. This might not much trouble us if the tendency were not enforced by temporal facts about action as well. People often say that action occurs only in the present, as if no one acted yesterday and none will act tomorrow. Obviously some untensed

---

[9] These remarks are more fully developed in Nerlich (1994: 269–72).

reading of the remark is meant, but which? The following facts may partly lie behind it.

My basic actions are done where and when I am: basic action is local. Of course, I can do a deal in London by phone from Adelaide, but the basic action of speaking into the phone is done here. I can write today so as to pay a bill by post in London next week. But the basic signing-the-cheque is local. Yet there are asymmetries in the ways basic action is temporally and spatially local. That springs, in part, from the facts about the action-guiding role of temporal perception; in part, from the spatial and temporal structures of common things.

Consider a basic action—a tennis forehand drive, say—not as a discrete act but as the continuously monitored movement described in coaching manuals. The core event of the stroke is the moment of racket–ball contact; let us call this event (we perceive the ball and racket contact as instantaneous) the *shot*. The *stroke* (the motion that includes the shot) is a developing process which must be begun, maintained, and followed through. The whole episode unfolds (let's say) within a *rally* and within a *court*. I play the shot where and when I am. We want to consider the time-skewed, local/non-local features in playing the drive which underlie the delusive intuitions of presentism. (There are surely more features than I shall describe.)

During the drive I should keep my eye on the ball. Let me try to analyse this in tenseless (and, for simplicity, absolutist) language.

Consider any of my times and positions within the drive but preceding the shot. (Ignore, for now, a shot where I must move to the ball.) At any such $\langle t_i, p_i \rangle$, I can see *where* the ball is (p_k) at t_i.[10] I cannot see, at $\langle t_i, p_i \rangle$, *when* the ball is at p_i. At $\langle t_i, p_i \rangle$ I can only judge and anticipate, not perceive it. The perception of the ball when it is at p_s occurs at t_s (the place and time of the shot) and at no other time. No perception shows me the ball at any time other than the time of that perception, but I can see it at any place (within the court) other than the place of that perception.

Two things are significant here (apart from the huge difference between the speeds of the ball and of light). The ball is temporally present throughout the shot, the drive, and the rally, but it is not spatially present throughout the court, nor is it spatially present to me during the drive, except at the shot. I am, myself, present throughout the rally but not everywhere in the court. For the sort of crudish perceptions at issue here, common things function like enduring, moving points, extended in time but not in space. My eye

---

[10] Or so it seems. *In fact*, I see the ball where it was an imperceptibly short time before the time of perception. The spatial difference is imperceptible, too. But my knowledge that this is so does not penetrate the perception, the seeing-as-if, even where the time-lag is quite perceptible, as in astronomical observations.

on the ball does not need to tell me when it is; it is throughout the rally always somewhere in the court. I have to see where it is *simultaneously with each moment* of my watching it. It is the (seeming) impossibility of my perceiving the ball at p_i at any time other than the time when it is at p_i which enforces the sense that perception and action occur instantaneously.

Second, at $\langle t_i, p_i \rangle$, I see that the ball is at p_k and also see it moving. Seeing something move and seeing that it moves differ. I see the second hand of my watch move; I cannot see the minute hand move, though I can certainly see *that* it moves. In each case, my having seen the hand opposite the numeral 5, say, affects my seeing it opposite 6.[11] But when I see the second hand move, the prior seeing affects the later seeing in a quite different way from the case of the minute hand. We do not know much about the difference between these causal processes, though we speak of short- and long-term memory. Do we have conscious access to these different processes? One way that could happen is by seeing a moving thing smeared. But that is unusual.[12] Nevertheless, it is an appealing thought that the seeing-things-move causal process is accessed via phenomenal cues such as smearing. But we need no such cues—we just see the ball moving. It is, then, a main factor in presentist delusion that often we do not see the moving ball as smeared. If it is perceived precisely as a ball then (it irresistibly seems) perceiving motion is instantaneous—takes no longer to occur than it takes for the ball to occupy that spherical volume. (Obviously, that cannot really be so.) This seemingly precise moment of perception seems also to be the moment of transition between times when information comes in and the latest time at which it can be used to guide racket to ball.

I have yet to say anything about action, but I think that this is easy. I swing the (temporally extended, spatially local) racket through the drive. This means pushing it, and the volitional structure of pushing is ineffably simple. I push differentially, adjusting speed and direction of hand and racket to my ever-changing perceptions of the ball. The volitional structure of that is ineffable, too. How the volitions are guided by perception can be described, but that task now turns out to be largely done.

Perceptual structures similar to those by which I keep my eye on the ball control my moving the racket head and my body parts. Of course, these are

---

[11] See Mellor (1981: 8). Seeing that *e* precedes *e'* means seeing *e* first. We perceive the order of events as the order of perceiving the events. The causal order—that seeing *e* affects one's seeing *e'*—fixes the perception of the order of events.

[12] More cautiously, smearing is not required for the perception of motion. I take this to be obvious for the case of slowish motion, which anyone may verify easily enough—just watch your finger move slowly across your line of vision. Unless you are more hawk-eyed than I am, you will see the motion but not the smearing. I find a wide range of slow speeds where the motion is obvious yet the outline sharp. But some things certainly move too fast for me (or, I guess, you) to see them clearly.

where I am at, throughout. However, my torso standardly functions as a rest frame for the purposes of actions which are direct bodily movements. I must perceive, relative to this, where head, eyes, arms, and legs are and how they are moving and, at the same time, I must move them into new positions and postures. These perceptions are kinaesthetic, proprioceptive, and less obvious to us as we use them. It is much harder for us to imagine, too, what it would be like to have no kinaesthetic perception of our body parts while retaining the capacity (blindly) to push. But the same structure of information flow is vital to this aspect of playing a successful drive.

I take it as obvious enough how these remarks extend to the role of temporal perception in actions quite generally.

The core, then, lies partly in delusions that states, and changes of state, are perceived instantaneously and simultaneously with the perceiving. Basic continuous action is a matter of moving oneself at each instant, simultaneously with perceiving oneself move. The (non-illusory) need to sustain attention continuously through the drive is reflected in the likelihood of a poor shot if I do not keep my eye on the ball until the moment that I play it. It is both the space-point/time-extended structure of the objects on which we act, the phenomenal structure of perceived motion and of the control of our own basic body actions which focuses us on the (seeming) instant of perception and performance. This syndrome is a mainspring of the delusion of presentism. Once identified, these aspects of perception and action are obvious and pervasive enough to make a significant step towards explaining the powerful hold which presentism has over us.

Tenseless thinkers have often agreed (with Smith, for example) that their theory does go against the grain of language somehow. Just how it does, what the consequences of that are (if any), and what needs to be done about them, are questions on which there is little agreement. I will not pursue those questions here. However, I think it likely that a spacetime metaphysics of time does place some burden of language reformation upon us, though that is another contentious issue. The most important conclusion of the arguments of this chapter is that we have deeply misunderstood the nature of time throughout most of human history. Though significant strides have been made in the sciences which point towards a new understanding, the metaphysical work of integrating these advances with the rest of our thinking is incomplete.

REFERENCES

Barnes, A. (1979), 'Cosmology of a Charged Universe', *Astrophysical Journal*, 227: 1–12.

Feinberg, G. (1967), 'Possibility of Faster-than-Light Signals', *Physical Review*, 159: 1089–1105.

Friedman, Michael (1983), *Foundation of Space-Time Theories: Relativistic Physics and Philosophy of Science* (Princeton: Princeton University Press).

Goldhaber, A., and Nieto, M. (1971), 'Terrestrial and Extraterrestrial Limits on the Photon Mass', *Reviews of Modern Physics*, 43: 277–95.

Mellor, D. H. (1981), *Real Time* (Cambridge: Cambridge University Press).

Nerlich, Graham (1994), *What Spacetime Explains: Metaphysical Essays on Space and Time* (Cambridge: Cambridge University Press).

Recami, E., and Mignani, R. (1974), 'Classical Theory of Tachyons', *Revista Nuovo Cimento*, 4 (series 2): 209–90.

Robb, A. (1914), *A Theory of Time and Space* (Cambridge: Cambridge University Press).

Smith, Quentin (1993), *Language and Time* (New York: Oxford University Press).

7

Absolute Simultaneity and the Infinity of Time

QUENTIN SMITH

1. Introduction: philosophy as a part of science

Philosophers frequently assume that the nature of things, events, properties, causality, free will, and the like is a matter for philosophers to determine, not scientists. But many philosophers assume that the nature of time is the topic of physics, specifically, Einstein's theory of relativity, and that this nature is not something that can be discovered in part or as a whole by philosophical investigation alone.

But this is not the case, if only for the reason that philosophy is a *part* of physics (and of the sciences in general); there is no such division between philosophy and science as is often supposed, or even the 'continuum' between philosophy and science that is associated with Quine-type views, which implies that at one end of the continuum there exists some philosophy that is not a part of science. For example, metaphysics and the philosophy of language can be viewed as the part of physics which is used to interpret the mathematical formulae or theorems that are taken to be confirmed by the observational data, and which is used to decide among the many theorems that are taken to be confirmed to some degree by the observational data.

Philosophy is a part of other sciences as well. According to the traditional *syntactic theory of science*, science consists of a system of theorems, taken as syntactic strings, rules relating these strings to possible observations, and semantic interpretations of these theorems. I would add that philosophy is the 'semantic interpretation' part of science. According to the more recent

I am grateful to a reader for Oxford University Press for helpful criticisms of a penultimate draft, which led to several changes. I am also indebted to Brian Leftow for extensive and influential comments on earlier versions of Sections 5 and 6 of this chapter, which led to numerous revisions. Comments offered by Craig Callender, William Lane Craig, Arthur Falk, Robin Le Poidevin, L. Nathan Oaklander, Michael Picard, and William F. Vallicella also proved very helpful and led to many revisions of the chapter. I should add that Oaklander has expressed disagreement with my critical analysis of his views (in Section 5), and I hope to deal more fully with his views in a future work. Similar caution should be expressed regarding my presentation of the Nerlich-inspired 'space analogy' argument in Section 5; this is an argument some of his unpublished writings suggested to me, so perhaps the 'suggestiveness', rather than the argument suggested, should be ascribed to him.

semantic theory of science, science consists of models and hypotheses about certain analogies and disanalogies of the models to reality. I would add that philosophy is the hypothesizing about the analogies and disanalogies of the models to reality.

With this understanding of philosophy, I will argue that the philosophical portions of the Special Theory of Relativity's concept of time, the General Theory of Relativity's concept of time, and the standard quantum mechanical concept of time consist in part of unsound verificationist arguments in metaphysics and the philosophy of language and that these theories need to be replaced by a certain non-verificationist theory I shall describe.

I shall argue that the correct, non-verificationist, philosophical part of physics implies that abstract objects exist in time, that temporal relations among all physical events are absolute rather than relative to a reference frame, that physical clocks do not measure the metric of time, that there is time before the big bang, that past and future time are infinite, and that time as a whole consists of an infinite number of infinitely long temporal series.

The present essay can be viewed as building upon the argument for the tensed theory of time articulated in Smith (1993a), specifically, the argument for the theory of 'absolute presentness'. This is a tensed theory of time that implies there is a single 'absolute tide of becoming' that encompasses all concrete and abstract objects. The theory of absolute presentness, rather than the theory of a tenseless, verificationist, and relativist time, is the philosophy of time that arguably belongs to the philosophical part of physics.

2. *Real time and the orthodox interpretation of quantum mechanics*

First in the order of things is to show that part of the philosophical theory of time in orthodox quantum mechanics, and Special and General Relativity, is an unsound verificationist argument. The verificationism of these theories has sometimes been noted in a general way in the philosophical literature,[1] but the verificationist arguments need to be pinpointed precisely and their unsoundness made plain to view.

Let us begin with the orthodox or Copenhagen interpretation of quantum mechanics. The orthodox interpretation of quantum mechanics has been accepted by virtually all physicists since the late 1920s. Most contemporary philosophers of quantum mechanics, and a small number of contemporary physicists, reject the Copenhagen interpretation; but here I am concerned to discuss the orthodox view that remains prevalent in the scientific community.

---

[1] The most extensive discussion of STR's verificationism appears in William Lane Craig (forthcoming).

The orthodox theory includes an *a prioristic* theory of time that is based on an invalid verificationist argument. The inclusion of this a prioristic theory was first evinced by Heisenberg's rejection of his original interpretation of the uncertainty relations between time and energy. Heisenberg originally interpreted these relations epistemically, but Bohr convinced him in private communications to accept a verificationist metaphysics, with its attendant ontological interpretation of the uncertainty relations; hence was born the orthodox interpretation of quantum mechanics. Although this interpretation originated mainly with some papers by Bohr and Heisenberg, the more speculative and somewhat idiosyncratic philosophical theories subsequently developed (in different ways) by Bohr and Heisenberg have not become part of the orthodox interpretation, but only a certain minimal component, a component that is found in textbooks on quantum mechanics (e.g. the typical textbooks used in graduate courses on quantum mechanics).

This component involves the idea that time cannot be measured to a precision greater than the so-called Planck time, 10^{-43} second, since the energy utilized by the physical clock process that is making the measurement precludes the possibility of a measurement of a shorter interval. A typical passage reads (Schlegel 1980: 77): 'Considering now the measurement of time, we give an indication that one cannot measure a time t for which ct is less than the Planck length [the Planck length is 10^{-33} cm]. Suppose a clock to have length dimensions of order L. If the clock is to measure a time interval t, the communication time L/c between its parts must be less than t.' The author proceeds to characterize the gravitational effect of the clock mass and the energy involved in the workings of the clock. 'Now, for a time measurement to be of significance the interval t must be greater than any alterations $\delta t'$ that are introduced by the gravitational effect of the clock mass and the energy ΔE utilized by the clock process . . . So the Planck length defines a theoretical limit on the smallness of a time interval' (1980: 78). Talk of intervals smaller than this length is meaningless. We have the inference:

> One cannot measure a time *t* for which *ct* is less than the
> Planck length. (1)

Therefore,

> 'There is a time for which *ct* is less than the Planck length' is
> not a true statement. (2)

This inference is based on a verificationist theory of the use of temporal words. Only if the concept expressed by 'a time' includes the concept of the means of verifying sentences about time does (2) follow from (1). But the verificationist theory of the meaning of words and sentences has long since been shown to be not only false but self-referentially incoherent. The principle of verification is meaningless by its own standards since there are no observations that could verify it. (See Smith 1997*a*: part I for a criticism of the argument that the principle of verification is an analytic truth. My criticism also implies that Quine's 'holistic verificationism', which retains the thesis that evidence-relations and semantic-relations are coextensive (Quine 1974: 38), is false.)

Since (2) does not follow from (1), there is no valid argument of this form that shows that 'time breaks down' at the Planck era near the big bang, namely, during what some describe as the first interval of 10^{-43} second after the big bang. Given this, the familiar line that 'QM implies that the concept of time ceases to be meaningful near the big bang' does nothing to show that time does not extend infinitely into the past.[2]

3. Real time and the Special and General Theories of Relativity

What is Einstein's justification for interpreting the variable t in the equations of the Special Theory of Relativity as referring to time? Although the a priori verificationist justification is present in his 1905 article, his most explicit articulation of this justification appears in his 1916 book *Relativity*. He asks us to consider

> all physical statements in which the conception 'simultaneous' plays a part. The concept does not exist for the physicist until he has the possibility of discovering whether or not it is fulfilled in the actual case. We thus require a definition of simultaneity such that this definition supplies us with the method by means of which, in the present case [where two lightning strokes hit near the front and back of a train], he can decide by experiment whether or not both the lightning strokes occurred simultaneously. *As long as this requirement is not satisfied, I allow myself to be deceived as a physicist (and of course the same applies if I am not a physicist), when I imagine that I am able to attach a meaning to the statement of simultaneity.* (See Einstein 1961: 22; my italics)

---

[2] It is clear that the orthodox interpretation of quantum mechanics includes an a priori verificationist metaphysics, but many new interpretations of quantum mechanics have emerged, especially in the 1980s and 1990s, and a different analysis of their metaphysical assumptions would be needed. (For example, see Smith 1997*b*.) But it is worth commenting that the Bohmian quantum cosmology developed by Callender and Weingard (1994) departs more radically than most other current physical theories from a verificationist stance. 'Like Newton's absolute time, the time in Bohmian cosmology is most naturally viewed as an unobservable, physical time, arising from the basic laws' (Callender and Weingard 1994: 224). They share Newton's belief that 'time was a real relation, not to be confused with its sensible measure' and that there is a 'preferred time' (ibid.). But whether or not they are committed to the false clock principle (8) discussed below is unclear.

Here we see Einstein defining 'simultaneity' in terms of a method of verifying statements in which this word occurs. Indeed, he says that apart from such verificationist definitions, 'simultaneity' and relevantly similar words are meaningless, both for the physicist and for everyone else. Clearly, this is an a priori philosophical theory and belongs to the philosophy of language and metaphysics. Einstein adopts the theory in the philosophy of language:

> The meaning of temporal words is a method of verifying the
> sentences in which these words occur. (3)

Note that Einstein provides no argument for this philosophy of language; he simply assumes it (as if it is obvious). These and other theories of Einstein show that the philosophy of language is indeed a part of physics. Thus, it is entirely appropriate to consider a theory of time that is based on a certain philosophy of language, such as the philosophy of language defended in Smith (1993*a*), to be a theory that is a part of physics (rather than some separate discipline that falls outside the realm of the sciences). One difference is that Einstein assumed, without argument, his philosophy of language, and drew inferences about time from these philosophical assumptions, whereas in Smith (1993*a*) I argued for a certain philosophy of language and drew inferences about time from this philosophy. There is no difference in kind between Smith (1993*a*) and Einstein's book *Relativity*; one difference in degree is that there is a greater number of mathematical equations in Einstein's book.

Einstein's formulation of the General Theory of Relativity is also based on a verificationist metaphysics. For example, he writes in his seminal 1916 paper 'The Foundation of the General Theory of Relativity':

That this requirement of general covariance [namely, that laws of motion are preserved by arbitrary transformations between any two coordinate systems], which takes away from space and time the last vestige of physical objectivity, is a natural one, will be seen from the following reflection. All our space-time verifications invariably amount to a determination of space-time coincidences. If, for example, events consisted merely in the motion of material points, then ultimately nothing would be observable but the meeting of the material points of our measuring instruments with other material points, coincidences between the hands of a clock and points on the clock dial, and observed point-events happening at the same place at the same time. The introduction of a system of references serves no other purpose than to facilitate the description of the totality of such coincidences. (See Einstein 1952: 117)

This passage suggests the inference:

> Verifications of statements about time are observations
> of coincidences between material particulars (e.g. marks
> on a clock dial and the hands of a clock). (4)

Therefore,

> Statements about time are statements about observable
> coincidences between material particulars.   (5)

'Time' is reductively definable in terms of an observable spatial relationship between one physical thing (the hands of a clock) and another physical thing (points on the clock dial).

With the advent of big bang cosmology, many cosmologists reductively defined General Relativistic time in terms of the expanding radius of the universe. For example, Christopher Isham states that the expanding universe solutions of Einstein's equations allow a slicing of 'spacetime' in which each three-space can be viewed as the three-dimensional spherical boundary of a four-dimensional ball. 'The time variable associated with this decomposition is the radius of the sphere . . . An absolutely crucial idea here is that "time" can be defined internally in terms of a particular property (i.e., the radius) of the curvature of the three-dimensional space' (see Isham 1988: 391).

Einstein came to accept a substantival theory of spacetime by 1920 and most philosophers of physics now accept a substantival interpretation of GTR and STR. Verificationism, however, also lies at the basis of these substantival interpretations, as I shall now argue.

4. Verificationism and substantival interpretations of relativity theory

It may be objected that my above criticisms pertain only to interpretations of STR, GTR, and QM that assume a reductionist and relational (Leibnizian) theory of time. If we adopt a substantival theory of time or spacetime, as many philosophers of physics have done since the 1970s, then (it may be alleged) my above criticisms of these theories do not hold.

I believe that the familiar substantival interpretations of STR, GTR, and the orthodox version of quantum mechanics are also based on a false verificationist metaphysics. I shall show this only for STR; it will be immediately apparent how my argument about STR applies also to GTR and the orthodox interpretation of QM.

Recently some philosophers who argue for a substantival theory of spacetime have argued that a non-verificationist formulation of STR is possible if we take it to be about (not luminal or causal relations but) spacetime, where spacetime is characterized in terms of Lorentz invariance. For example, Nerlich (1994: 251) says that 'Einstein's early infatuation with positivism, and the ensuing darkness which this shed on the deep riches of his discoveries is an unhappy chapter in the brilliant tale of science in this century.' Philosophers such as Nerlich argue that Minkowski spacetime is a substantival entity and that the constant speed of light is not a matter of convention.

However, this substantival and non-conventionalist interpretation of STR still rests on a verificationist metaphysics. This theory requires that the variable *t* in the Lorentz transformations and *s* in the Minkowski formula for spacetime separation refers, under certain conditions, to time.[3]

But what justifies the assumption that, given a substantival and non-conventionalist theory of STR or Minkowski spacetime, *t* or *s* refers in suitable cases to a time or temporal interval?

A perspective on this question can be obtained by comparing STR with the theory it 'replaced', Lorentz's theory. Note that Lorentz himself did not interpret the *t* and *t'* in the Lorentz transformation equations as referring to times, since he took time to be the referent of a variable in the Galilean transformation equation. In Lorentz's theory, the Lorentz transformations have two steps, the first step giving us the Galilean coordinates (of events in absolute space and time) and the second step giving us 'apparent' spatial and temporal coordinates, which is what measuring instruments record in frames that are either moving or at rest in absolute space. (Lorentz called these apparent times 'local times'.) The Galilean transformations give us the values that time and space have relative to an inertial frame at rest in absolute space. The Galilean coordinates are the ontologically real coordinates of events in time and space. The electromagnetic laws hold only in the inertial frame at rest in absolute space, even though they appear to hold in all inertial frames (they appear to be covariant, i.e. to have the same form in all inertial frames). But it is not physically possible to verify which inertial frame is at rest in absolute space, so we cannot know the relevant values in the Galilean transformation. All that can be observed are the values in the second step of the Lorentz transformation, which give us the measurable or apparent spatial and temporal values in a given inertial frame. (In one frame, the one at rest in absolute space, the apparent coordinates are also the real coordinates, but we cannot know which frame this is.)

Now the Special Theory of Relativity, be it formulated in terms of a relationalist and conventionalist metaphysics, or in terms of a substantivalist and non-conventionalist metaphysics, entails that nothing real corresponds to the Galilean transformation, that there are no Galilean coordinates, and that the apparent coordinates are the only ontologically real coordinates. This

---

[3] For example, Penrose writes that the length *s* in the spacetime distance formula,

$$s^2 = t^2 - (x/c)^2 - (y/c)^2 - (z/c)^2$$

is a 'time interval' in cases where an observer is moving in our chosen coordinate system, and that *t* refers to 'time' for clocks at rest in our chosen coordinate system. See Penrose (1989: 196).

is entailed by the Principle of Relativity and the Light Principle, for if the electromagnetic laws really hold in all inertial frames and light really has the constant velocity c in all inertial frames, it follows that the apparent coordinates are the only real coordinates.

What justifies these two principles of STR? The Lorentz theory and STR are observationally equivalent (cf. Zahar 1989); furthermore, Lorentz's theory is not ad hoc, as has recently come to be recognized (Zahar 1989, Craig forthcoming). In light of such facts as these, what justifies omitting the Galilean transformation?

The standard or 'official' justification offered by contemporary defenders of substantival interpretations of STR is the appeal to simplicity. It is said that it is simpler to assume that the only temporal relations, time-series, etc. that exist are the ones postulated by the substantivalist formulation of STR, than it is to assume that there also are other, absolute temporal relations, an absolute time-series, for which there is no observational evidence.

But this standard justification is undermined by the fact that substantival STR is in fact a much less simple theory than the theory of absolute time. Substantival STR postulates infinitely many real time-series in addition to the observable physical clock processes. (It postulates one Minkowski spacetime, but infinitely many time-series, one for each set of inertial frames that are mutually at rest.) But the theory of absolute time postulates only one real time-series in addition to these clock processes. The theory of absolute time implies the Lorentzian 'local times' are merely apparent times; that is, they are mere appearances, and are not temporal series that belong to the furniture of reality. There exist real clocks, but the 'local times' these clocks may be said to 'measure' do not in fact exist. All that exists is one real time-series, absolute time.

The STR substantivalist and the defender of absolute time both posit (i) indefinitely many observable physical clocks, but the STR substantivalist in addition postulates (ii) indefinitely or infinitely many real time-series, whereas the defender of absolute time instead postulates (iii) only one real time-series. This is reason to believe that the theory of absolute time is simpler and more ontologically economical than substantival STR.

Furthermore, STR posits infinitely many 'contents' of the laws of nature, specifically, of the mechanical and electromagnetic laws, one for each set of inertial frames that are at rest relatively to each other, even though these contents have the same 'form' in each frame. The Lorentz theory, however, postulates only one content and one form, the content and form the laws possess in the inertial frame at rest in absolute space. In Maxwell's electromagnetic laws, E = electric field and H = magnetic field. In a Lorentzian theory, these laws have one content; Maxwell's equations have the content $[E, H; x, y, z, t]$ in absolute space and time, where x, y, and z are the spatial coordinates and t the temporal coordinate. But in STR, these laws have a different content in each co-moving inertial frame; in one frame there obtain the laws [E, H; x, y, z, t], in a relatively moving frame there obtain [E', H'; x', y', z', t'], and so on *ad infinitum*. In STR, the laws have the same form in each frame (are covariant), but the infinite multiplicity of different contents makes its postulation of laws infinitely less simple than the Lorentzian postulate of one content.

A widely accepted view today is Zahar's 'conspiracy of silence' argument against the Lorentzian theory, developed at greatest length in his (1989). In an earlier essay (1983) he gives a clear encapsulation of the argument:

One could argue as follows: it is unlikely that Nature contains both deep asymmetries and compensatory factors which exactly nullify these asymmetries. Such a state of affairs is not logically impossible and to envisage it is not meaningless; but it is unlikely, or improbable, in the same intuitive sense in which a series of coincidences and accidents having a single global effect are improbable. (1983: 39)

This argument may be questioned. A series of coincidences or accidents having a single global effect are compounds of nomologically independent events that have a single global effect. A *coincidence* is inexplicable and the compound of events constituting the coincidence has necessary and/or sufficient conditions (or probabilistic conditions) that are independent of one another (see Owens 1992). However, in a Lorentzian theory, what Zahar calls the 'compensatory factors which exactly nullify these asymmetries' *are* nomologically explained and have the same sufficient and necessary conditions (or have conditions that are not independent of one another). There is an ether in absolute space that transmits both the electromagnetic forces and molecular forces. This transmission causally affects (is a causally sufficient condition of) various properties of the forces. Lorentz deduces the Lorentz–Fitzgerald contraction hypothesis, time dilation, and the *measured* constancy of the velocity of light from this transmission causal hypothesis. The Galilean transformation gives real temporal and spatial coordinates which, due to the time dilations and length contractions that are derived from the transmission causal hypothesis, appear as the measured 'local' or 'apparent' coordinates that are governed by the Lorentz transformations. This is exactly the opposite of 'a series of coincidences and accidents' that have a single global effect.

In fact, the real crux of Zahar's objection to the Lorentzian theory is a verificationist assumption, despite Zahar's extensive denials that verificationism is a premiss of his critique of Lorentz's theory. He writes: 'But there is something paradoxical in that, through postulating a universal medium, we are driven to conclude that it must be *undetectable*. Was it not dissatisfaction with this paradox so closely connected with the crucial experiment which caused Einstein to look for another explanation?' (Zahar 1989: 92, my emphasis). Zahar himself, in arguing that Lorentz's theory is not ad hoc in any sense, maintained that Lorentz causally explained the apparent time dilations and length contractions by his 'molecular forces hypothesis' (Zahar 1989: 50–1 and 74), contradicting Zahar's assertion that Lorentz's theory is analogous to postulating a series of coincidences that have a single global effect. To resolve this contradiction in Zahar's theory, we have to eliminate the 'series of coincidences' thesis from Zahar's formulation of the 'conspiracy of silence' criticism, which leaves as a remainder merely the thesis that an undetectable entity is paradoxical. It is only a verificationist assumption that could justify the thesis that the *undetectability* of a universal medium is a *sufficient reason* for concluding that its postulation is *paradoxical*. The hypothesis 'x exists and is undetectable' is paradoxical *by itself* if and only if it is true a priori that 'whatever exists is detectable'.

I suggest that, faced with this rebuttal of the 'argument against absolute time based on simplicity or the "series of coincidences" objection', STR substantivalists would none the less feel theoretically reluctant to abandon STR in favour of the theory of absolute time. I suggest, further, that the reason for their reluctance is a deep but *implicit* commitment to certain a priori theses in metaphysics and the philosophy of language. (I do not mean they would say or admit this, but that this is the ultimate justification that is behind— whether they recognize it or not—their allegiance to STR.) They share with the relational interpreters of STR the fundamental ontological decision to eliminate 'physical unobservables and immeasurables', i.e. to take 'the real' and 'the physically measurable or verifiable' as expressing logically equivalent concepts.

One main difference between the STR relationalist and the STR substantivalist is that the substantivalist does not *identify* time with observable physical clock processes, but rather takes observable physical clocks as giving us the accurate measurements of time. The substantivalist assumes that statements about time, even though they are not *about* observable clocks, are none the less about something that is accurately measurable by observable clocks. STR substantivalists and relationalists are implicitly committed to the a priori theses:

> Necessarily, sentences about the topological and metrical
> properties of time are verified or falsified by, and only by,
> the observable properties and relations of physical clocks;   (6)

and

> Necessarily, the topology and metric of time are accurately
> measurable by observable physical clocks.   (7)

Thesis (6) belongs to the philosophy of language and (7) belongs to meta-physics. (Clocks allegedly measure the topology of time at least inasmuch as they allegedly measure the order in which events or times occur; the metric involves the temporal distance between events or times.)

If the reader supposes that 'clock' analytically expresses a concept of an instrument that accurately measures the metric of time, we should substitute a more neutral expression for 'clock', such as an 'apparently periodic phys-ical process' or 'a type of motion of a physical particular between spatial points that is apparently regularly repeated'. The question then would be whether such apparently regular motions are clock processes. But I shall assume 'clocks' refers only to such moving physical systems and does not analytically entail that they accurately measure the metric of time. I take 'x is accurately mea-surable by observable physical clocks' to entail 'x is knowable by observing the clocks'.

Let us consider the metaphysical thesis (7), which is weaker than (6) since it asserts a sufficient condition ('by'), not both a sufficient and necessary condition ('by and only by'). The necessity operator is not that of physical or nomological necessity, since (7) is used as a criterion to determine what could be physically necessary. There are many metaphysically possible worlds (or what Plantinga 1974 and Forbes 1985 call 'broadly logically possible worlds') in which material bodies can be accelerated to arbitrarily high vel-ocities (not limited by a finite velocity of light) and in which the operand of (7) is true. (The operand of (7) is expressed by 'the topology and metric of time are accurately measurable by observable physical clocks'.) These worlds are physically impossible, according to STR. This is a consequence of the fact that the necessity operator in (7) does not have a merely physical modality (its truth does not depend on the truth of STR), but a metaphysical modal-ity. (7) is supposed to be a true proposition that belongs to several different physical theories (the true physical theory and several false ones) and that determines which physical theories are *possibly* true. This metaphysical assumption is at the basis of both the relational and substantival interpreta-tions of STR and is required to rule out the Galilean transformations.

But this metaphysical assumption, like the principle of verification, is not true. There is a metaphysically possible world in which the Lorentz theory is true; in this world, the structure of time is given in the Galilean transforma-tions, but the values of these transformations cannot be known by observing any clock. But we need not rely on Lorentz's theory to show this; any world in which time has a topology and metric that cannot be known by observations of physical clocks (but in which Lorentz's laws also do not obtain) is a world that shows (7) is false. (6) is also false, for observations of physical clocks would neither verify nor falsify statements about the metrical and topological properties of time in these worlds.

(6) and (7) are false a priori metaphysical assumptions that are a part of STR and that are used to interpret the observational data provided by physical clocks. Without these verificationist metaphysical assumptions, STR collapses; there is no longer any justification for accepting the Principle of Relativity and the Light Principle. The observational data require us to accept the hypotheses that the laws of electrodynamics (and mechanics) *appear* to hold (with same form) relative to every inertial frame, and light *appears* to have a constant velocity c, independent of the state of motion of the emitting body, relative to every inertial frame. But the passage from 'seeming' to 'being' is made through an a priori verificationist theory of 'being'.

In order to develop an adequate theory of time, we need to reject the a priori propositions (6) and (7) and consider a relevant a posteriori proposition. The proposition that is relevant to determining the nature of time a posteriori is:

> The topology and metric of time are accurately measurable by
> observable physical clocks. (8)

The truth-value of (8) is knowable via the correct metaphysical part of physics in conjunction with the observational data relevant to the mathematical equations in physics.

In the following several sections, I shall present some arguments that imply that (8) is false. First, I need to show that abstract objects exist in time (Sections 5 and 6). Second, I will show (in Section 7) that it follows by a series of steps that all simultaneity relations between physical events are absolute (are two-termed relations) and are not relative to a reference frame (three-termed relations). This indicates that proposition (8) is false, for observable physical clocks do not record absolute temporal relations between distant physical events. These conclusions enable us to argue that *time probably preceded the big bang*, that the past and future are probably *infinite*, and that time probably consists of an *infinite sequence of infinitely long temporal intervals* (see Section 8).

5. Necessary and sufficient conditions for an object to exist in time

My argument for absolute simultaneity and the infinity of past and future time hinges on the admittedly controversial assumption that there are some abstract objects whose existence is not dependent on the existence of any concrete object. There exist universals that are not exemplified, there exist propositions even if no one is thinking of them, and there are numbers even if there are no minds or bodies. In other words, I assume platonic realism.

For a large number of philosophers, platonic realism is a preposterous theory they cannot imagine believing. But it is also true that a large number of philosophers find anti-platonism, especially in its nominalist version, a preposterous theory they cannot imagine believing. (Anti-platonism includes Aristotelian realism, conceptualism, physicalism, trope theory, and the many varieties of nominalism.) The debate between platonists and anti-platonists has been going on for over 2,000 years without any sign of a 'knock-down argument' or a consensus of opinion among philosophers in sight. However, I do not want to make my argument of interest only to platonists.

The largest group of philosophers who find platonism distasteful are physicalists. To make platonism seem more palatable to physicalists, I would point out that most physicalists, such as Quine, have argued that physicalism requires some abstract objects, sets (to which numbers arguably may be reduced), since the postulation of sets is required by the physical sciences. If Quine is right, then even physicalists are committed to the existence of abstract objects. Quine plausibly argued that the physical sciences require sets to be postulated, but Tooley (1987) plausibly argued in a similar spirit that the physical sciences require uninstantiated properties to be postulated. Given this, I do not see that platonism can be so unpalatable to physicalists. I do not see much difference between believing in the existence of the unit set of Socrates or the null set or the infinite hierarchy of power sets of sets, and holding the platonist belief that there is an uninstantiated property of *having 867 sides*.

In any case, my entire argument below, given in terms of platonic realism, is logically equivalent to an analogous argument that does not assume platonic realism but is about merely the null set, or about unit sets, or ordered *n*-tuples of physical particulars, or sets of sets. Thus, physicalists such as Quine would be able to accept the basic premiss of my argument without abandoning their philosophical scruples. I shall briefly present one argument for absolute simultaneity that is based solely on physicalism, with its sets, after I have presented the detailed argument in terms of platonic realism.

First, we need to give a real or true (as opposed to stipulative) definition of an object's existence in time. A stipulative definition is neither true nor false and merely records one's decision to use a word in a certain way; for example, one may stipulate that one will use 'God' to refer to the universe, so that *x* is God if and only if *x* is the universe. We may object to this stipulative definition that it does not facilitate communication but makes it more difficult, but we cannot object that it is false. By contrast, a real definition purports to describe the nature or essence of something. If we want to give a real definition of *being human*, we will specify the logically necessary and sufficient conditions for something to have the nature of a human. If we offer the real definition, *something is human if and only if it is a rational animal*, then this definition is either true or false. This definition is false if some humans are not rational (e.g. are born and live without a cerebral cortex) or if some humans are not animals (e.g. have their animal parts gradually replaced by bionic parts) or if some rational animals are not humans (e.g. whales).

In the following real definitions of an object's existence in time, the variable x ranges over concrete and abstract objects, but not over instantaneous events or enduring processes. By 'n-adic property' I mean a monadic property or a relation of any sort. The crucial aspect of these definitions is that the second-order variable F ranges over real and Cambridge n-adic properties.

> x exists in time if and only if there is some time t at which
> x possesses some n-adic property F and some different time
> t' at which x does not possess F. (D1)

The tensed version is as follows:

> x exists in time if and only if x now possesses F and either did
> not or will not possess F; or x will possess F and either does
> not now possess F or did not possess F; or x used to possess
> F, and either now does not possess F or will not possess F.
> (The 'or' expresses an inclusive disjunction, meaning and/or.) (D2)

My subsequent arguments that all abstract objects exist in time, that simultaneity and presentness are absolute, and that the past and future consist of infinite sequences of infinitely long temporal intervals, all hinge upon the fact that (D1) and (D2) are true and, specifically, upon the fact that Cambridge properties are included in the domain over which the second-order variable F ranges. If there are no Cambridge properties and no Cambridge changes, then abstract objects do not exist in time and my arguments for absolute simultaneity, for infinite time, and for the falsity of the 'clock principle' (8) presupposed by STR, GTR, and QM are unsound arguments. Thus, I shall first refute a series of arguments purporting to show that Cambridge changes and properties do not exist. After this, I will give some positive arguments that Cambridge changes and properties exist. I will then show that it is logically necessary that something exists in time if it undergoes a Cambridge change.

First, I would note that the 'real/Cambridge change' terminology is unfortunate, since it conversationally implicates (in Grice's sense) that Cambridge changes are not 'real' in the sense of not existing. Moreover, the real/Cambridge distinction is both vague and equivocal and is often used to mark several different distinctions. For example, one distinction may be called the 'internal change/external change' distinction, which is the difference between a change in *the parts or monadic properties of an object x* and a change in *y*'s relation to the object *x* that is due only to a change in the parts or monadic properties of *x*. However, I shall use the 'real/Cambridge' terminology since (even though an adequate definition is hard to come by) most philosophers can intuitively recognize what is a Cambridge change and what is a real change.

Do there exist Cambridge changes and Cambridge properties, such as the null set's relational property of *being apprehended by John* or a proposition's relational property of *being believed by Jane*? Consider one argument for the non-existence of these properties.

There are some propositions about related items that change their truth-value entirely due to changes in one of the objects, without implying the other relational term has any monadic property in virtue of which the truth-value changes. For example, if a proposition *p* is believed by Jane at 12.00 a.m. and is not believed by Jane at 12.01, then a different proposition, *p is occurrently believed by Jane*, changes its truth-value entirely due to changes in Jane's mental states. There is no monadic property of *p* by virtue of which the proposition *p is occurrently believed by Jane* undergoes a change in truth-value. One can (allegedly) infer from this that *p* does not undergo any existent change or acquire or lose any property and thus that Jane's occurrent belief does not give any reason to think that *p* exists in time.

However, this argument (and each logically related argument for the non-existence of Cambridge changes) is flawed for a number of reasons.

(*a*) The inference that the proposition *p* does not acquire or lose any property, and thus that there is no reason to think *p* exists in time, is invalid. All that follows is that *p* does not acquire or lose a monadic property. The premisses of the argument are consistent with the thesis that *p* acquires and loses the relational property of being believed by Jane, that this is an existent Cambridge change that *p* undergoes, and that *p* thereby exists in time.

(*b*) An additional problem with this argument is that the proposition *p is occurrently believed by Jane* does not change its truth-value by virtue of a monadic property of either Jane or *p*. The proposition changes its truth-value by virtue of a dyadic property being transiently exemplified by Jane, namely, the dyadic property of *occurrently believing p*. However, this is logically equivalent to the fact that the proposition *p is occurrently believed by Jane* changes its truth-value by virtue of *p* transiently exemplifying the dyadic property of *being occurrently believed by Jane*. Consequently, this account gives no logical distinction between real and Cambridge changes.

Further, this doctrine is self-contradictory, since it says that the proposition *p is occurrently believed by Jane* changes its truth-value; but a proposition changing its truth-value is a paradigmatic instance of a Cambridge change, which supposedly is not an existent change at all.

(*c*) Perhaps the anti-Cambridge theorist can formulate an argument that does not include a premiss that implies a certain proposition undergoes a Cambridge change. The anti-Cambridge theorist may say that acquiring and losing a relational property is not a change at all (a change that exists), and therefore that there are no Cambridge changes. But this statement is provably false, since if a change is defined as the acquisition or loss of a monadic property, the absurd result follows that the people at Hiroshima *being blown up by an atomic bomb* is not a change they underwent, since *being blown up by a bomb* is a relational property. In fact, the paradigmatic instances of changes are acquisitions or losses of relational properties, e.g. being killed by, being injured by, being broken by, being reshaped by, moving from x to y, etc. These changes cannot all be reductively analysed into changes of monadic properties, if only for the reason that movement or change of place consists primitively in relations of the moving object to other objects or places.

(*d*) There is another strategy for denying that Cambridge changes exist. One can define changes (i.e. all changes that exist) in terms of acquiring or losing an *n*-adic property *by virtue of a causal event or process*. But this strategy for eliminating Cambridge changes is logically untenable for one of two reasons.

First, it implies that it is logically possible that changes in mental states are not changes. If agent causality and the libertarian theory of free will are true, such that mental events are freely caused by the agent, then the mental event E of *believing the proposition p* will be caused by the agent A, but the more complex event of A *causing* E will not itself have a cause. If the next mental event E' that A experiences is also caused by A, then the event, A *causing* E, is neither an effect of anything nor a cause of anything. (The brain events would be parallel or subvenient to the mental events but not causally related to them.) It would follow that the agent's causing her mental event E and then causing her mental event E' is not a mental change. But this is absurd, since such changes of mental state are paradigmatic instances of changes.

Second, we may argue that Cambridge changes satisfy this causal definition of a change. (This argument does not depend on libertarianism or the theory of agent causality.) A mental belief-event E does *bring about* something, the instantiation of an *n*-adic property; it brings about the instantiation of such a property as *being believed by Jane*. Since the proposition p acquires this property as a result of the event E occurring, p's acquisition of this property is due to a causal event occurring, the event E.

It may be objected that p's acquisition of this property is not an effect of the event E since p is an abstract object and thus cannot be affected by anything.

However, it is false that abstract objects cannot be affected by anything. Even though Quine is a concrete object, any singular or *de re* proposition that includes him as a part is an abstract object. The *de re* proposition consisting of Quine and the property of being alive, {Quine, being alive}, is affected if one of its parts is affected. If Quine is blown up by a bomb, then this *de re* proposition is affected; indeed, the explosion causes this proposition to cease to exist, since the impact destroys one of the parts of the proposition (Quine). This singular proposition ceases to exist in the sense that it exists at (occupies the time) t but does not exist at (does not occupy) the later time t'.

There is a difference between two types of effects, abstract effects (which are types of Cambridge changes) and concrete effects. This does not imply Neoplatonism is true, since both abstract effects and concrete effects can be brought about by concrete causes, such that there are no abstract causes of concrete events. But some abstract effects have abstract causes; a *de re* proposition's ceasing to exist is an abstract event that causes my *de dicto* belief that this proposition exists to undergo the abstract event of becoming false.

(Of course, this account of the causal definition of change implies that my first argument against the causal definition of change (the argument about the complex event, *agent A freely causing E*) does not tell against the causal definition of a change; it tells against this definition only if all causes and effects are concrete occurrences.)

(*e*) A fallacy of equivocation is one reason many philosophers believe that Cambridge changes do not exist, a fallacy that is tempting due to the conversational implicature of 'real change'. They believe 'x undergoes a merely Cambridge change' does not imply 'there is some change that x undergoes'. But the relevant argument (x underwent a change c that is not real; therefore, c does not exist) is based on a fallacy of equivocation on 'real'; in the premiss it means something such as 'a change in a concrete thing's parts or monadic properties' and in deriving the conclusion it is taken to mean instead 'a change that exists'.

Let us now pass to a positive argument that Cambridge changes exist. It can be argued that 'c is a Cambridge change and c does not exist' is implicitly self-contradictory. A familiar example of a Cambridge change is *being remembered* and then *not being remembered*. If I remember my dead grandmother at noon, 24 August 1996, and cease to remember her by 12.01, 24 August 1996, my grandmother has undergone a Cambridge change. The familiar line is that the only change that exists is a change in my mental states or myself; I have changed from the mental state of remembering her to the mental state of not remembering her. I acquired and lost a property, but my grandmother did not, or so the familiar line goes.

But the familiar line is implicitly self-contradictory. If my grandmother did not acquire and later lose *any* property, then she did not acquire and lose the *n*-adic property of *being remembered by me*. She did not stand to me in the relation of being remembered, and then cease to stand to me in this relation. One of two possibilities occurred instead: (i) my memory image stands to me in the relation of being remembered, and then no longer stands to me in this relation, or (ii) my act of remembering is a monadic property, such that nothing stands to me in the relation of being remembered.

But the first option is self-contradictory, since I did not remember my *memory image*, but my grandmother. My memory image exists simultaneously with my occurrent mental act of remembering, but the relation of remembering (by definition) is a relation to some thing or event that exists earlier than the act of remembering. At 12.05 I can remember the memory image I formed at 12.00, but this is different from remembering my grandmother walking across her lawn in 1970.

The second option, that my act of remembering is a monadic property, is also self-contradictory. If it is a monadic property, then it is an intentional act that has no intentional object. It is a conscious act but there is nothing of which it is conscious. It is not 'about' anything. However, remembering, by definition, has an intentional object, namely something remembered. The proposition *there occurred an act of remembering in which nothing was remembered* is analytically false. Of course, I can seem to remember something and, due to a defect in my epistemic faculty, not remember anything. But I am here talking about *rememberings* (i.e. *seeming rememberings* that are true seemings and thereby are *rememberings* or acts of remembering).

Each intentional act, by virtue of having an intentional object, does not require the *extra-mental* existence of the intentional object. If I fantasize a unicorn walking on a tightrope, the intentional object is a fantasized unicorn, and '*x* is a fantasized unicorn' implies '*x* exists dependently on the act of fantasizing'.

The only option left is the third one, that *my grandmother* stands to me in the relation of being remembered, and later does not stand to me in this relation. Since 'standing in a relation' implies 'exemplifies an *n*-adic property', it logically follows that my grandmother exemplified an *n*-adic property and at a later time did not exemplify it. My grandmother underwent this Cambridge change and therefore there is a change my grandmother underwent. A logically analogous argument can be used to show that there is some *n*-adic property acquired and lost in each case of Cambridge change, and that in each such case there exists a Cambridge change.

So far, I have argued that Cambridge properties exist and that my definitions (D1) and (D2) of existing in time cannot be refuted by showing that there are no Cambridge properties. However, the rebuttal of these arguments does not imply that (D1) and (D2) are true. For example, one could concede that Cambridge changes exist but still hold that temporal existence should be defined in terms of real changes. Thus, I need some positive arguments that show (D1) and (D2), with their intended implication that undergoing Cambridge changes is logically sufficient for existence in time, are true definitions of temporal existence.

According to (D1) and (D2), it is irrelevant *what sort* of *n*-adic property is acquired or lost (be it real or Cambridge), since it is logically impossible for an object to acquire at one time some *n*-adic property and lose at a later time that property if that object exists timelessly. '*x* exists timelessly and yet possesses different *n*-adic properties at different times' is an implicit logical contradiction. I shall demonstrate this by first considering some arguments that this is not a logical contradiction.

In William F. Vallicella's interesting article 'No Time for Propositions' (1995), he argues that there is no reason to accept a certain premiss of one of my arguments in Smith (1990), namely, that 'Something is in time if it stands in a relation at a time.' Vallicella writes: 'suppose proposition p stands to John at time t in the relation of being believed. If propositions are timeless, then [Smith's premiss that something is in time if it stands in a relation at a time] is false' (Vallicella 1995: 474). Vallicella concludes that my argument is question-begging.

I did not offer a justification for this premiss since it seemed obvious to me, but a justification can be offered. If *p* stands in *R* to *x* at time *t*, it follows that what is expressed by the adverbial modifier 'at time *t*' modifies what is expressed by 'stands in'. In other words, *being at t* is a property of *p*'s *standing in the relation R*, and if something exemplifies an *n*-adic property at a certain time, then that object exists in time.

Vallicella says that *p*'s *standing in relation R* exists in time, but it does not follow that *p* exists in time. Vallicella suggests that I commit a fallacy of division. Consider the whole event, John's occurrently believing that *p*. This event occurs in time. But the fact that this whole belief-event occurs in time does not imply that each part of this belief-event, such as the proposition *p*, occurs in time.

Vallicella notes that the Cambridge belief-event of *p's standing in a belief-relation to John* is an event that is simultaneous with other events; this is 'an event and therefore the kind of thing that can be simultaneous with other events, and thus unproblematically in time. But p, although it stands in the belief-relation, is not an event' (Vallicella 1995: 474). Vallicella concludes that my argument that *p* exists in time is unsound.

If Vallicella's argument is that only events exist in time, that *p* is not an event, and therefore that *p* does not exist in time, we can dismiss his argument out of hand, for his theory would imply that persisting objects, continuants, do not exist in time. But surely electrons, monkeys, and galaxies exist in time.

Perhaps his argument may be more charitably construed as the argument that some objects exist in time, but that 'an object O's *exemplification of an n-adic property F* is an event simultaneous with other events' does not imply 'the object O exists in time'. The reason there is no implication, the argument may go, is that there are Cambridge properties and *exemplifying Cambridge properties simultaneously with other events* is not a logically sufficient condition for an object to exist in time. An object must transiently exemplify a *real n*-adic property in order to exist in time.

This suggests that Vallicella's argument in effect reduces to the second main argument against my definitions (D1) and (D2). The first main argument is that Cambridge properties do not exist. The second main argument is that Cambridge properties exist but that objects exist in time only if they acquire or lose some real property.

However, this is no argument at all but merely a question-begging assertion. There are no discernible premises to 'the argument' that an object must undergo a real change in order to exist in time. This assertion seems instead to be a stipulative definition of 'exists in time' or a linguistic convention that most philosophers and scientists have tacitly agreed to adopt. But stipulative definitions and linguistic conventions are non-alethic items and thus cannot threaten the thesis that (D1) and (D2) are true definitions of time's nature.

Suppose, however, that some objectors assert that *it is true* that objects have to undergo real changes to exist in time, and claim that my contrary assertion (namely that (D1) and (D2) are true) is no more or less a question-begging assertion than their own assertion. The objectors may be content with a 'stand-off' with me, which would allow them to maintain their 'real change' definition without fear of refutation.

However, I shall show that the thesis that Cambridge properties exist, and yet that temporal existence is defined in terms of real properties, is inconsistent with relevance logic. Our objectors believe that the sentence:

> p stands in a belief-relation at time t_1 and does not stand in
> this relation at the later time t_2 (9)

does not entail

> p exists in time.   (10)

But they claim that:

> x stands in a causal relation at time t_1 and does not stand in
> this relation at the later time t_2 (11)

entails

> *x* exists in time. (12)

However, if (11) entails (12), in the sense of relevance logic, then (9) entails (10). The parts of the sentence that are relevant to the entailment are exhibited in the following argument:

> *x* stands in an *R*-relation at time t_1 and does not stand in this
> relation at the later time t_2; (13)

therefore,

> *x* exists in time. (14)

The word 'causal' is not the word that makes the premiss (11) relevant to the conclusion (12), namely '*x* exists in time'. Rather, it is the temporal words in the premiss, 'at time t_1' and 'at the later time t_2'.

Note, for example, that there is no relevant entailment in this argument:

> God stands in a causal relation to abstract objects; (15)

therefore,

> God exists in time. (16)

There is no part of (15) that enables it relevantly to entail (16). To provide such an entailment, an extra premiss is needed, such as 'whatever stands in a causal relation exists in time'. (Even if causes and effects are only events, and never objects, an object can stand in a causal relation at one time and not at another time in the sense that it exemplifies a causal property, or undergoes a change that is a causal event, at one time but not at a later time.)

Some philosophers argue that the open sentence '*x* causes *y*' strictly implies '*x* is earlier than *y*'; other philosophers argue that '*x* causes *y*' strictly implies '*x* is simultaneous with *y*'; still others argue that it implies '*x* is earlier than or simultaneous with *y*'; and still others that it implies '*x* is either earlier than, simultaneous with, or later than (backwards causation) *y*'. And other philosophers argue '*x* causes *y*' strictly implies '*x* is earlier than, simultaneous with, or later than *y*, or *x* exists timelessly and *y* exists in time, or *x* and *y* exist timelessly'. Perhaps one of these views is correct. But the correctness of any such view would do nothing to impugn my argument that '*x* causes *y*' does not entail in the sense of relevance logic that '*x* exists in time'. If *p* strictly implies *q*, then it is logically necessary that *p* materially implies *q*. The proposition *q* is true in every possible world in which *q* is true. But this falls short of entailment in the sense of relevance logic. For there to be such entailment, an *n*-adic predicate in the conclusion must appear in at least one of the premisses and in the present case '( ) exists in time' or 'is temporal' appears in the conclusion but not in the premiss, which features instead the *n*-adic predicate 'causes'. Because there is no relevant entailment of conclusions about time from premisses about causes, there can be an open and debatable question about which temporal or atemporal conclusions (if any) are strictly implied by premisses about causal relations.

Similar arguments hold for other real *n*-adic properties in terms of which temporal existence may (allegedly) be defined, such as light-connectibility, movement, and the like. In short, the thesis that Cambridge properties exist but that temporal existence is defined in terms of real properties is not consistent with relevance logic.

It is interesting that Vallicella (1995) concedes that some propositions exist in 'wide time', but not the 'narrow time' of the concrete causal order. This concession is all I need for my definitions of existence in time to be true, for '*x* exists in wide time' entails '*x* exists in time'. The adjectives 'wide' and 'narrow' refer to non-temporal properties of the items that exist in time. '*x* exists in narrow time' means '*x* exists in time and is a concrete particular that bears causal relations to other concrete particulars'. '*x* exists in wide time' means '*x* exists in time and need not be a concrete particular that bears causal relations to other concrete particulars'.

Note that '*x* exists in time' in my definitions and in the conclusions of my arguments does not entail '*x* endures from the first time (at which it possesses a Cambridge property *F*) to the later time (at which it does not possess *F*)'. If we are talking about a permanently existing *de dicto* proposition or null set, we are entitled to infer that it remains present from the first time (when I come to be aware of it) to the second time (when I cease to be aware of it). In terms of the tenseless theory of time, we are entitled to infer that it is temporally located at the same time, t_1, that my first act of awareness is located, is temporally located at the same time, t_2, at which the event of my ceasing to be aware of the abstract object is located, and that the object is located at all the times in between t_1 and t_2. But if a past object, such as my dead grandmother, is first remembered and later not remembered, it does not follow that my grandmother is present from the first time to the second time. It does follow that my grandmother *existed at some times* throughout the time period from the first time (the time of my remembering) to the second time (the time of my not remembering). My grandmother existed at times with different degrees of pastness; that is, she became more and more past from the first time to the second time. In terms of the tenseless theory of time, my dead grandmother has a greater temporal distance from the second time, 12.01 a.m., than she does from the first time, 12.00 a.m.; specifically, she is earlier than 12.01 a.m. by one more minute than she is from 12.00 a.m.

There may be a different sort of objection to (D1) and (D2) from the sorts I considered in the preceding paragraphs. It may be objected that my definition of temporal existence, if true, would imply, by parallel reasoning, that the definition of spatial existence would imply the absurd conclusion that abstract objects exist *in space*. An objection of this sort was suggested to me by Graham Nerlich, in connection with my argument in Smith (1993a: 209–10) that abstract objects exist in time. I argued that if some abstract object possesses at one time t_1 the property of being referred to by John and does not possess this property at the later time t_2, then the abstract object acquires and loses the property of being referred to by John and thus exists in time.

A parallel argument of the sort Nerlich suggested concerns 'exists in space'. If some abstract object O possesses the property of *being referred to by John* when John is at place P, and loses this property when John moves to another place P', then O acquires and loses the property of *being referred to by John* when John is at different places and thus O exists in space.

I would respond that this argument is not parallel, since there is no relevant analogy. The thesis that the abstract object O exists in time is not inferred from the premiss that *John* exists at different times when he refers and does not refer to O. Rather, it is inferred from the premiss that John's *referring to O* exists at one time and not at another time. But John's *referring to O* does not exist at any place (even though John exists at different places). John's mental act of referring to O lasts for two seconds, but is not extended through any space (it does not have a certain height, width, and depth) and is not a mass-point that is located at any place.

6. All abstract objects exist in time

(a) First argument

So far I have argued that if a proposition stands in a belief-relation to somebody, that proposition exists in time. This is not sufficient to show that all propositions exist in time, for many propositions are not believed by anybody. However, if I have a belief *about* all propositions, e.g. that all propositions are non-spatial, then all propositions exist in time, for in this case all propositions would acquire and lose some property.

Vallicella believes this argument is invalid. He writes:

Now surely the following argument is invalid:

> I believe that every proposition is nonspatial
> q is a proposition

Therefore

> I believe that q is nonspatial. (1995: 477)

Vallicella notes that 'my thinking doesn't "reach" q due to the opacity of belief contexts' (1995: 477).

Vallicella believes that my argument only shows that the general proposition *every proposition is non-spatial* acquires and loses a belief-relation to me. This is true; however, each proposition, such as *q*, acquires and loses a different relation, not to me, but to my occurrent act of believing the general proposition; each proposition acquires and then loses a truth-making relation to my occurrent believing. The proposition *q*, and each other proposition, transiently stands to my occurrent believing in the relation of *making it true*. After I no longer hold this belief, *q* and each other proposition no longer stand in this truth-making relation to my belief. Thus, my argument about believing a general proposition about all propositions does, in fact, show that all propositions exist in time.

Of course I am not saying that *q* is *the* truth-maker of my belief in the proposition *every proposition is non-spatial*. Rather, the sort of truth-making relation in which it transiently stands to my occurrent believing is *being a necessary part of the truth-maker of* the believing. The truth-maker of my occurrent belief is the conjunctive state of affairs whose conjuncts are *q is non-spatial, r is non-spatial, . . .*, and so on for each proposition. This state of affairs is also the truth-maker of the proposition *every proposition is non-spatial*. This conjunction permanently stands in a truth-making relation to the proposition and transiently stands in a truth-making relation to my occurrent believing.

(b) Second argument

Vallicella notes that my argument regarding beliefs shows only that propositions exist in time contingently; they exist in time if there happen to exist intelligent organisms that believe propositions. However, I think there are more general arguments that show that propositions, and all other abstract objects, exist in time if time exists (regardless of whether or not intelligent organisms exist).

For each event *e*, when *e* is present, there is a true tensed proposition, *e is present*, that corresponds to the state of affairs composed of *e*'s being present. When *e* becomes past, then this proposition no longer has the property of corresponding to any state of affairs, since there no longer is the state of affairs of *e*'s being present. The proposition acquires and then loses the *n*-adic property of being true and thus the proposition exists in time. The same holds for propositions of the form *e is past by two seconds*, or *e is future by two seconds*, or any other transiently true proposition. The same holds for each transiently exemplified universal; e.g. the universal *three-sidedness* is first exemplified by a sand formation on the beach and is later not exemplified by it.

A related argument shows that each non-transiently truth-valued proposition and each unexemplified universal exists in time. Each non-transiently truth-valued proposition p is a conjunct of a more complex proposition p and q, such that q transiently corresponds to a state of affairs. This implies that p acquires and loses the property of being a part of a conjunctive proposition one of whose conjuncts is transiently corresponding to a state of affairs. Further, each unexemplified universal is a part of some such proposition p and thus acquires and loses the property of being part of a conjunctive proposition one of whose conjuncts is transiently corresponding to a state of affairs.

(c) Third argument

There is a more general argument that shows that any object, concrete or abstract, exists in time if time exists. If time exists, then there is some temporal state of affairs, *the presentness of the instant x*, that is presently different from each concrete and abstract object. When x is present, each concrete and abstract object has the corresponding relational property of being presently different from the obtaining state of affairs, *the presentness of x*. When x becomes past, each concrete and abstract object loses the relational property of being presently different from (what was) the obtaining state of affairs, the *presentness of x*; they do not now stand in the relation of being different from x's *being present*, since there now is no state of affairs consisting of x's being present.

(d) Fourth argument

L. Nathan Oaklander believes (private communication) that my arguments show only that if the tensed theory of time is true, all abstract objects exist in time. Oaklander argues that my tenseless definition (D1) does not show that three-sidedness exists in time. The tenseless definition is:

> x exists in time if and only if there is some time t at which x
> possesses some n-adic property F and some different time t'
> at which x does not possess F. (D1)

Oaklander argues that universals do not exist in time. He argues that if a sand formation on the beach is three-sided at t_1 and is not three-sided at t_2, it does not follow that *three-sidedness* acquires and loses the property of being exemplified by the sand formation. He states that the sentence

Three-sidedness is exemplified by the sand formation on the
beach at time t_1 (17)

is ambiguous between (18) and (19):

At t_1, three-sidedness exemplifies *being exemplified by the
sand formation on the beach*;   (18)

Three-sidedness (timelessly) exemplifies *being exemplified
by the same formation on the beach at t_1*.   (19)

Oaklander states that (18) implies that three-sidedness exists in time, but that
(19) does not imply this, and that my argument does nothing to show that
(19) is false.

I would respond that the insertion of '(timelessly)' in (19) makes the argument question-begging, since Oaklander would then be assuming rather than
proving that three-sidedness exists timelessly. Sentence (19) is supposed to
be the conclusion of an argument offered by the defender of timeless properties,
not a premiss of the argument.

If we remove '(timelessly)' from (19), then we may grant that (19) is a
valid interpretation of the sentence (17). However, there is no difference in
meaning between (18) and (19). In both cases, the adverb 'at t_1' modifies the
verb 'exemplifies'. The adverb 'at t_1' is a temporal adverb; it tells us when
the occurrence mentioned by the verb takes place. Whether this adverbial
modifier is placed at the beginning or end of the sentence affects the syntactic form of the sentence, but not its semantic content. (18) and (19) have
the same meaning; they express the same proposition.

If Oaklander objects that 'exemplifies being exemplified by . . .' does
not express a property, (18) and (19) may be rephrased. For example, (19)
becomes:

Three-sidedness characterizes the sand formation on the beach
at t_1.   (20)

Here again, 'at t_1' adverbially modifies the verb 'characterizes' and tells us
when this characterization occurs. Accordingly, it is not necessary to adopt
a tensed theory of time to show that abstract objects exist in time. The conjunction of the tenseless definition (D1) and (20) implies that three-sidedness
exists in time (given the premiss that three-sidedness does not characterize
the sand formation at t_2).

It might be objected that one can view 'at t_1' in (20) not as an adverb mod-
ifying 'characterizes' but as an adjective modifier that functions to pick out
a particular temporal slice S_1 of the sand formation of the beach, where the
time-slice S_1 is the temporal part of the sand formation that is temporally
located at t_1. We may then read (20) as

Three-sidedness belongs to the S_1 time-slice of the sand
formation on the beach. (20′)

It may then be argued that three-sidedness does not belong to the time-slice
of the sand formation at any particular time, but timelessly.

One problem with this argument is that it presupposes that the substance
or continuant theory of objects is false. If the substance theory is true, the
sand formation is not a string of time-slices or temporal parts; rather, the sand
formation is an enduring substance that has a history and its history has temporal parts. The temporal part of the sand formation's history that is located
at t_1 is not three-sided; rather, the sand formation is three-sided, and its exemplification of this property belongs to the temporal part of the sand formation's history that is located at t_1.

The critic may well respond that she may be presupposing the bundle theory of objects, but that this is legitimate in the present context since I am
presupposing the continuant or substance theory of objects and have no more
'refuted' the bundle theory than she has 'refuted' the substance theory. At
the very least, the critic may argue that there is no sound argument for the
temporality of universals if the bundle theory of objects is true.

The effective response to this critic is that the bundle theory no less than
the substance theory of objects entails that three-sidedness exists in time. Three-
sidedness belongs to time-slice S_1 in the sense of being a part of the bundle
of properties that constitute the sand formation at t_1. The S_1-slice of the sand
formation is a bundle of monadic properties that are located at t_1. Since three-
sidedness is a part of S_1, three-sidedness is a part of a bundle of properties
located at t_1. If a bundle of monadic properties is located at t_1, each monadic
property that is a part of this bundle is located at t_1. (This argument is not a
'fallacy of division' since it is based on the nature of bundles of monadic
properties.) It follows that three-sidedness is located at t_1 and thus that three-
sidedness exists in time.

If abstract objects exist in time, this shows a metaphysical assumption of
STR, GTR, and orthodox QM, namely,

The topology and metric of time are accurately measurable by
observable physical clocks, (8)

can be shown to be false. I will argue from the premiss that abstract objects
exist in time to the conclusion that physical events have relations of absolute simultaneity to other, distant, physical events. If simultaneity is absolute, then the observational data indicate that observable clocks do not
accurately record these relations and that (8) is false. This is argued in the
next section.

7. *Physical events are absolutely temporally related to each other*

If all abstract objects exist in time, then at some time t, all the abstract objects that exist at t bear a two-termed (absolute) relation of simultaneity to each other. For example, right now, at time t, propositions p and q are both present, and sustain a relation of simultaneity with each other. It must be an absolute relation of simultaneity, since there is no spatial midpoint between the two propositions at which light or other physical signals can arrive simultaneously relative to a reference frame.

Physicalists, by virtue of postulating sets, are also committed to absolute temporal relations among sets, and (I shall argue) to absolute temporal relations among all physical events.

But before I discuss physicalism and sets, I shall state my main argument in terms of propositions. If propositions exist in time, they (like any other concrete or abstract objects or continuants) have a history, and this history is a sequence of temporal parts. What does it mean to say that a proposition has a history, and that this history is a whole of temporal parts? Each temporal part of a proposition p's history is earlier or later than other parts. One temporal part of p's history may include p's standing in the relation of correspondence to a transient state of affairs S. If p is an intransiently true proposition, such as *triangles have three-sidedness*, a temporal part may include p's being composed of a universal (e.g. three-sidedness) that is being transiently exemplified by something x. If p is intransiently false and contains some universals that are never exemplified by concreta, p may have a temporal part that includes p's transiently standing in a certain relation to a transiently false proposition q, namely the relation of p and q both being presently false. Numerous other such relational properties are included in the temporal parts of a proposition's history.

These considerations enable us to present an argument for absolute simultaneity among some distant physical events.[4]

To begin with, let us consider a proposition q that we shall suppose is presently true, namely, that *Socrates died and Arnap is born*. Let 'Arnap' be the name of some organism in the distant galaxy Andromeda, such that Arnap's birth is not related to the event of Socrates' death by the relations denoted in STR by 'topologically earlier than' or 'topologically later than' (which either are, or are logically equivalent to, relations of *x's being causally connectible to y by signals at or less than the velocity of light*). According to STR, no *absolute* relation of earlier, later, or simultaneous connects Socrates'

---

[4] This argument for the absolute simultaneity of distant physical events is additional to the two arguments I presented in (1993*a*: 239–41).

death to Arnap's birth. (But STR implies these two events are connected by a two-termed or absolute relation denoted by 'topological simultaneity', which is equivalent to the relation of *x's being causally unconnectible to y by signals at or less than the velocity of light.*[5]) But we shall see that absolute B-relations (earlier, later, or simultaneous) do connect distant events.

The fact that the proposition *q*, *Socrates died and Arnap is born*, is presently true does not imply that the death and birth are both simultaneous or are both present; indeed, Socrates' death is past (we shall suppose we do not know which temporal relation Arnap's birth has to Socrates' death; thus, we treat 'Arnap is born' as tenseless). But the death and birth are presently standing in some type of partial correspondence or truth-making relation to the presently true proposition *q* (this is entailed by the two premises that *q* is presently true and that truth is correspondence). Socrates' death and Arnap's birth each correspond to different parts of the *de dicto* proposition *q* (the parts are different complexes of interrelated universals that partly compose the proposition *q*); each of these two concrete events partly composes the state of affairs that is the truth-maker of *q*. The death's and birth's *corresponding to a part of q* are abstract events or states of the death and birth; that is, they are abstract events involving the concrete events of the dying of Socrates or the being born of Arnap. (I use 'state' or 'event' to refer to something's exemplification of an *n*-adic property, where the variable 'thing' has an unrestricted range.)

Perhaps it is worth emphasizing that the concrete event of Socrates' death is not the same event or state as the abstract state of Socrates' death standing in a certain type of truth-making relation to a presently true proposition. One event consists in Socrates' exemplification of dying and the other event consists in this first event's exemplifying *corresponding to a part of the proposition q*.

The two present, abstract events or states are absolutely simultaneous with each other. Why? The proposition *q* presently stands in a correspondence relation to the state of affairs consisting of Socrates having died and Arnap being born and this state of affairs presently stands in a correspondence relation to *q*. Each of the two concrete events (Socrates' death and Arnap's birth) that make up this state of affairs presently stands in a partial correspondence relation to *q*. The abstract event or state of presently *corresponding to a part of q* exists in time but not in space. The relation of *corresponding to a part of q* does not have height, width, or depth and is not a mass-point. Something spatial (e.g. air molecules or neutrinos) that touched every part of the places

---

[5] Nerlich and others reject the tradition of Robb and others and assert that relations of topological earlier etc. are not definable in causal terms but are instead truth-makers of subjunctive conditional statements about causal connectibility.

at which Arnap's birth and Socrates' death occur would not touch or come into spatial contact with the birth's and death's *corresponding to a part of q*.[6]

Given this, there is no spatial midpoint between the abstract events of Socrates' death presently *corresponding to a part of q* and Arnap's birth presently *corresponding to a part of q*. Thus, there is no spatial midpoint between these two abstract events at which light or other physical signals could arrive simultaneously relative to a physical reference frame. The B-relation (simultaneous, earlier, or later) between the two abstract states is not relative to a reference frame, but is absolute or two-termed.

The next step in the argument for the absolute B-relatedness of distant physical events involves the fact that the concrete events of Socrates' death and Arnap's birth are temporally related to the abstract events of their presently corresponding to a part of *q*. This is an absolute B-relation, since there is no spatial midpoint between the event of *Socrates' death* and the non-spatial event of Socrates' death *corresponding to a part of q* that is required for a frame-relative B-relation. Socrates' death is past and thus is absolutely earlier than the present abstract event of his death's standing in a correspondence relation to a part of *q*.

Consider the temporal part *t* of the history of the proposition *q* that consists of *q* now (at noon, 28 July 1996) standing in a relation of correspondence to the state of affairs. This present temporal part *t* is absolutely simultaneous with the two abstract events consisting of the death's and birth's now corresponding to a part of *q*. Since absolute B-relations among events are transitive, the present temporal part *t* is also absolutely B-related to the death and birth. If the death and birth are absolutely B-related to the present temporal part *t* of the proposition's history, they are absolutely B-related to each other.

This argument is based on the transitivity of absolute B-relations among events. If Socrates' death is absolutely B-related to the abstract event of Socrates' death presently *corresponding to a part of q*, and this abstract event is absolutely B-related to the temporal part *t* of *q*, and *t* is absolutely B-related to the abstract event of Arnap's birth presently *corresponding to a part of q*, and this latter abstract event is absolutely B-related to Arnap's birth, then Socrates' death is absolutely B-related to Arnap's birth. Since Socrates' death and Arnap's birth are distant events, we are not in a position to know which

---

[6] Talk of 'partial correspondence to *q*' or 'corresponding to a part of *q*' may be misleading terminology. The relations of parts of propositions to parts of states of affairs are the relations of identity, replacement, or instantiation (see Smith 1993*a*: 151–8); correspondence is the relation of the whole proposition to the whole state of affairs. However, it would take up too much space to explain this theory of the correspondence relation and so I shall use my present terminology.

absolute B-relation obtains between them. But some absolute B-relation obtains between them, and hence the theory of time in STR is false.

This argument may be applied to any proposition that defenders of STR mistakenly believe entails that simultaneity is relative. Consider this *de dicto* proposition:

> Two distant events E_1 and E_2 emit light signals that are observed
> to arrive simultaneously at the apparent spatial midpoint between
> E_1 and E_2, relative to the reference frame R, but the signal from
> E_1 is observed to arrive first at the apparent midpoint relative to
> the reference frame R'. (p)

Defenders of STR infer from this proposition that the B-relation between E_1 and E_2 is frame-relative; but it can be shown that this proposition instead implies that E_1 is absolutely B-related to E_2.

Suppose proposition p is presently true, i.e. presently corresponds to the physical state of affairs S described by p. This entails that S presently stands to p in the relation of corresponding to p. (It does not entail, of course, that S or any of its parts is present; items that are past or future can presently stand in correspondence relations to presently true propositions, or at least so I argued in Smith 1993a.) Since there is no spatial midpoint between S and p and no spatial place relative to which the correspondence relation obtains, the present event or state of p's standing in the correspondence relation to S is *absolutely* present.

Each part of the state of affairs S, such as the concrete events E_1 and E_2, presently stands in a correspondence relation to a part of the presently true proposition p. E_1's present standing in this relation is an abstract, present state of E_1. Each such abstract, present state of a part of the state of affairs S may be called an abstract state A. Each abstract state A is absolutely B-related to the part of S of which it is a state (since there is no spatial midpoint between A and the part that could make the B-relation relative to a reference-frame). For example, the concrete event E_1 is absolutely B-related to the abstract state A_1 that consists of E_1 presently corresponding to a part of p. Each present, abstract state A cannot be present relatively to a reference frame, or be simultaneous with the other abstract states relatively to a reference frame, since there is no spatial midpoint between these abstract states; these abstract states are instead present and simultaneous absolutely.

Each abstract, present state A is absolutely B-related to such physical parts of S as E_1 and E_2. Since absolute B-relations among states or events are transitive, it follows that E_1 is absolutely B-related to its abstract state A_1, that A_1 is absolutely B-related to the abstract state A_2 of E_2, that A_2 is absolutely B-related to E_2, and therefore that E_1 is absolutely B-related to E_2. Thus, the two distant physical events E_1 and E_2 are absolutely B-related. Consequently, the proposition p implies that B-relations among distant physical events are absolute, not frame-relative.

This, of course, is consistent with the light signals emitted from E_1 and E_2 appearing to arrive at the apparent midpoints between E_1 and E_2 at different times, relative to different reference frames; but this is a fact about how light signals appear to observers (or would appear to observers, if observers were present), not a fact about time.

Thus, we have the irony that the very propositions that state STR's theory of relative B-relations, such as the proposition p, entail that B-relations are absolute.

It might be denied that absolute B-relations among events are transitive, and therefore that a crucial premiss of my above argument does not succeed.

One possible reason for denying this transitivity is that absolute B-relations are based on spatial relations. It might be said there is a possible world in which the spatial universe splits into two spatially unconnected parts, and then joins up. In this world, A is earlier than B, which is earlier than C. And D is earlier than E, which is earlier than F. And A is absolutely simultaneous with D, and C is absolutely simultaneous with F. If the spatial universe splits in two and later joins up, it will not follow from the above temporal relations that *either B is earlier than E, or B is simultaneous with E, or B is later than E.* For the event B could occur in one spatially disconnected part of the universe and the event E could occur in another spatially disconnected part, in which case B and E would not stand in a temporal relation. This would show that absolute B-relations need not be transitive.

In response to this argument I would note that my previous arguments have shown that non-spatial, abstract objects or events (such as propositions or the histories or temporal parts of propositions) are absolutely B-related and that physical and spatial events are transitively B-related by means of being B-related to the temporal parts of these propositions' histories. The physical event B is absolutely simultaneous with the temporal part p_1 of the proposition p's history and the physical event E is absolutely simultaneous with this same temporal part p_1 of the proposition p's history, and thus B is absolutely simultaneous with E. Absolute temporal relations are not based on spatial relations; they are logically and metaphysically independent of spatial relations and thus there is no difficulty in two spatially disconnected events B and E being absolutely B-related.

It can be reasonably thought that such relations as *absolute simultaneity* are intransitive if one is talking about relations between continuants (objects, substances) and other continuants or events. A substance or continuant O may at one time be simultaneous with an event E_1 and at a later time simultaneous with an event E_2, such that E_2 is not simultaneous with E_1. Absolute B-relations are also intransitive if we are talking about B-relations between processes of different temporal lengths.

However, if we are talking about relations between *instantaneous* events or temporal parts of objects (or events that endure for the *same* amount of time), then absolute simultaneity is transitive. The above example can be ana-lysed by saying that one instantaneous temporal part t_1 of the history of the continuant O is simultaneous with the instantaneous event E_1 and a later instan-taneous temporal part t_2 of the continuant's history is simultaneous with the instantaneous event E_2.

Other relations that the word 'simultaneity' can be used to designate need not be transitive. For example, the relation of *spacelike separation* discussed in the Special Theory of Relativity is often called the relation of 'topological simultaneity' and this relation is not transitive. But whether the relation of spacelike separation is transitive is a logically independent issue from whether or not the relation of absolute simultaneity is transitive.

The role that transitivity plays in the argument for absolute simultaneity can be further clarified if we evaluate Richard Swinburne's criticism of one of the two arguments for the absolute simultaneity of distant physical events I presented in (1993*a*), the argument that appealed to mental events. Swinburne has pronounced this argument a 'disaster' (1996: 489). How-ever, a closer inspection shows that the calamity has instead occurred in Swinburne's counter-argument, which we may now take the opportunity to witness.

Swinburne quotes a sentence from page 239 of *Language and Time*: 'An intelligent organism near the star Arcturus may be thinking of the proposition that *a triangle is three-sided*, and I may be thinking of it, as well, such that both these acts of thinking are absolutely simultaneous with the proposition and thereby with each other.'

Swinburne comments: 'But of course "thereby" is only warranted if the proposition exists only at one instant, and no one holds that; Smith holds that propositions exist sempiternally' (1996: 488–9). Swinburne has in mind the fact that since the proposition exists at all times, the intelligent organism's act of thinking of the proposition may occur in (say) 1815 and then be abso-lutely simultaneous with the proposition, and my act of thinking may occur in 1996 and then be absolutely simultaneous with the proposition; thus there is no reason to think these two acts of thinking are absolutely simultaneous with each other. (Even if Swinburne were right in his analysis of my sentence, which he is not, the conclusion still follows that the two acts of thinking are absolutely B-related to each other, which is all that is needed for my argu-ment for absolute time.)

But Swinburne is not right, since he commits a scope-fallacy; he assumes, quite arbitrarily, that the relation signified by 'thereby' does not lie within the scope of the possibility operator ('may be'), even though the placement of 'thereby' in the sentence implies it lies within its scope. The sentence in question does not mean:

(Sw): It is possible that both myself and some intelligent organism near the star Arcturus are thinking of the proposition that a triangle is three-sided; it is also possible that my act of thinking is absolutely simultaneous with the proposition and that the other organism's act of thinking is absolutely simultaneous with the proposition; if our two acts of thinking are absolutely simultaneous with the proposition, they are absolutely simultaneous with each other.

Here the inference to the absolute simultaneity of the two acts of think-ing is placed outside of a possibility operator. However, my sentence that Swinburne quotes includes the 'thereby' within the scope of the possibility operator; the sentence means:

(Sm): It is possible that both myself and some intelligent organism near the star Arcturus are thinking of the proposition that a triangle is three-sided; it is also possible that my act of thinking is absolutely simultaneous with the proposition and that the other organisms's act of thinking is absolutely simultaneous with the proposition; it is in addition possible that we are thinking of the proposition absolutely simultaneously.

(It then can be argued, as I do in Smith 1993a: 239, that it is possible that our two mental acts are absolutely simultaneous with our respective brain events and that our respective brain events are absolutely simultaneous with each other; given this, the possible situation envisaged is one where our respective brain events are two distant physical events that are absolutely simultaneous.)

Swinburne's counter-argument is like saying that 'It is possible that Jane pushes the rock and that the rock thereby rolls down the hill' is a false sen-tence since 'Jane pushes the rock' does not imply 'the rock rolls down the hill'. But the quoted sentence embodies a valid argument, since the consequence relation falls within the scope of the possibility operator; the possible situation envisaged is one where the consequence relation (in this case a causal relation) obtains. Swinburne would have done better to respond to my criticism of his own theory of absolute simultaneity (Swinburne 1981: 202), namely, that Swin-burne mistakenly identified Newton's theory of absolute time with the theory of cosmic time in the General Theory of Relativity (see Smith 1993a: 245).

We need not appeal to the fact that propositions (as distinct from the tem-poral parts of their histories) bear simultaneity relations to concrete events. Using the above example, we can legitimately refer instead to temporal parts, so that the relevant sentence reads: 'An intelligent organism near the star Arcturus may be thinking of the proposition that a triangle is three-sided, and I may be thinking of it, as well, such that both these acts of thinking are absolutely simultaneous with the same temporal part of the proposition's history and thereby with each other.' This sentence is logically equivalent to a sentence that is similar except for the fact that the 'thereby' is placed outside of the scope of the possibility operator.

Physicalists may reject my arguments for absolute presentness and absolute simultaneity since my arguments appeal to propositions. But similar arguments may be given that involve only distant physical events and their absolute B-relations to *sets*, which are the only abstract objects standardly postulated by physicalists. I have argued in Sections 5 and 6 that sets, like other abstract objects, exist in time and thus have temporal relations to each other. Given this, there is an absolute B-relation between Socrates' death and the unit set whose member is Socrates' death. There is no spatial midpoint between Socrates' death and the set of which his death is the sole member that could make a relative B-relation possible. There is also an absolute B-relation between the unit set of Socrates' death and the unit set of Arnap's birth in the distant galaxy Andromeda, since there is no spatial midpoint between these two unit sets that could make a relative B-relation possible. (Light signals do not travel back and forth between sets; a set does not have height, width, or depth, and is not a mass-point, and thus is not located in space. If they are not located in space, there can be no spatial midpoint between them.)

Since Arnap's birth is absolutely B-related to the unit set of Arnap's birth and this unit set is absolutely B-related to the unit set of Socrates' death, which in turn is absolutely B-related to Socrates' death, it follows that Socrates' death is absolutely B-related to Arnap's birth. Thus even an anti-platonic physicalism implies that STR is false.

What does my theory of absolute time share in common with STR?

Alan Sidelle (1995) criticized my argument for absolute simultaneity in Smith (1993*a*) by saying that if time is absolute, but some light relations or other causal relations are frame-relative, problems arise about time travel and closed causal loops. Sidelle insightfully notes that my critique of STR in *Language and Time* is not a critique of STR's verificationism, but is a different sort of criticism (my critique of STR's verificationism appears only as one of the parts of my argument against STR in the present chapter). Sidelle makes the observation:

He [Smith] argues that the special theory is not, as it seems, a theory about time, but about, for instance, causal relations between luminal signals. While not without problems, his argument does not amount to charges of verificationism, and is rather novel. One question is whether Smith's interpretation commits him to backwards causation. If the (possible) causal relation between events is relative to a frame of reference, but there is nonetheless an absolute temporal ordering of events, then relative to one of these frames of reference, some event can causally influence another that is absolutely earlier. (See Sidelle 1995: 680)

I would make two observations. First, in Smith (1993*a*) I argued that my theory of metaphysical time shares with STR the thesis that the (possible) causal relations between physical events are *not* relative to a frame of reference, but are absolute and frame-invariant; the STR phrases 'absolutely earlier' (or 'topologically earlier') and 'absolutely later' (or 'topologically later') denote these possible causal relations. There is an absolute and frame-invariant causal ordering of physical events in both STR and my theory.

Second, what I argued to be frame-relative is neither causal relations between non-luminal physical events nor causal relations between light signals, but the observable spatial midpoints between physical events (see Smith 1993*a*: 243). These two theses are not consistent with the possibility that, relative to one reference frame, some event can causally influence another that is absolutely earlier in metaphysical time.

The frame-relativity I endorse, which is the frame-relativity of observable spatial midpoints, implies that the topological temporal order and metrical relations among causally unconnectible physical events are not accurately measurable and knowable by observable physical clocks. But the temporal topological order (times as ordered by B-relations) of causally connectible events is accurately measurable by observable physical clocks.

The frame-relativity of observable spatial midpoints also leads us to the thesis that metrical relations, or lengths of temporal intervals, between any two successive physical events are not accurately measurable by observable physical clocks. The reason for this is that the observable spatial midpoints between the relevant clock-events are frame-relative. For example, if I take myself to be at rest, then the midpoint between the event of a light signal S leaving the bottom mirror of my light clock and the event of its arriving at the top mirror will appear to be located on a line perpendicular to the bottom mirror. If a light signal S is sent from the bottom to the top mirror, S will pass through this apparent midpoint, at a place we will call P. But someone in motion relative to me, who takes herself to be at rest, will observe the spatial midpoint through which S passes to be at a different place P', which lies along the direction of motion in which she takes me to be moving. This is because my light clock will appear to her to move in this direction, and thus the light signal (in its journey to the top mirror) will appear to have a longer distance to travel to reach the top mirror. My light signal and her light signal both appear to her to have the same velocity c. Since my light signal appears to her to traverse a greater distance than the signal in her own similar light clock (which she takes to be at rest), she will observe my light clock to be running slow, e.g. to measure a time less than an hour during the time that her light clock measures one hour. This apparent clock retardation is reciprocal; if I take myself to be at rest, her clock appears to me to be running slow.

Since neither I nor the person in the co-moving frame can determine who is really at rest and who is really moving, we cannot know which light clock accurately measures the lengths of temporal intervals. For this reason, the metric of time is not accurately measured by observable clocks.

We can infer from this the falsity of the principle

> The topology and metric of time are accurately measurable by
> observable physical clocks. (8)

Notice that we did not derive the falsity of (8) from an argument that a Lorentzian or neo-Lorentzian physical theory is true. Rather, we derived the falsity of (8) from our argument for absolute B-relations among all abstract and concrete events, in conjunction with the observational statement that the observable spatial midpoint between physical events is relative to a reference frame. Thus, we reject the 'received wisdom' that the following conditionals are true: (*a*) 'if STR is false, a neo-Lorentzian theory is true', (*b*) 'if time appears frame-relative but is really absolute, a neo-Lorentzian theory is true'.

I believe the argument from the temporality of abstract objects to the absoluteness of temporal relations among all physical existents, and the theory of absolute presentness in Smith (1993*a*), are the metaphysical core that ought to replace the verificationist, tenseless, and relativist theory of time that is the metaphysical core of STR and GTR. (The tenseless theorist may adopt just the part of my metaphysical theory that is consistent with the tenseless theory of time, namely the anti-verificationism and the absoluteness of B-relations.)

If we apply my realist or anti-verificationist metaphysical core to cosmological considerations, what shall we conclude? One conclusion is that there is time before the big bang, that past and future time are infinite, and that time consists of an infinite series of infinitely long temporal series. This is argued in the following section.

8. Infinitely many infinite temporal series

Since orthodox QM and GTR, with their verificationist core, are not about time, the fact that orthodox QM and GTR break down at the Planck epoch or the big bang (i.e. their laws are inapplicable) gives no reason to think that time begins then. But does the realist metaphysics of time give us positive reason to think that time preceded the big bang?

Suppose that concrete objects began to exist at the big bang. Is it none the less possible that time existed prior to the big bang? This is possible if only for the reason that abstract objects could exist earlier than the big bang. Some abstract objects, such as *de dicto* propositions and purely qualitative universals, exist non-dependently on concrete objects, and they can exist in time by virtue of undergoing Cambridge changes. For example, the proposition *there presently are concrete objects and real changes* could be false prior to the big bang and successively acquire and lose Cambridge properties such as *will be true in three seconds, will be true in two seconds*, etc. For there to be time before the existence of concreta, it suffices that there be some abstract object that undergoes some Cambridge change, such that this Cambridge change does not require some real change to occur concurrently with it or occur earlier than it.

Furthermore, since the occurrence of some Cambridge changes is independent of any concurrent real changes, the essential properties of these changes, including their topological and metrical properties, are not dependent on any concurrent real changes. For example, some mathematical proposition's property of *being true for a year* does not require that the earth revolve once around the sun; it entails merely that the proposition is true for an interval of time that would be occupied by the earth revolving once around the sun, if there were to exist such an earth and sun. Such counterfactuals enable us to talk about hours, years, etc. prior to the big bang, even if there exist only abstract objects before the big bang. The truth-makers of these counterfactuals are some possible worlds that are the most similar in the relevant respects to the actual world.

But we need not think that it is logically or metaphysically necessary that concrete objects or real changes exist *at some time* in order for there to be time. If there were no concrete objects ever, there could still be time. (We may follow Newton-Smith and call times 'quasi-abstract objects' or 'quasi-concrete objects', since they are relevantly dissimilar from both abstract and concrete objects.) It is sufficient that there be some abstract object that undergoes some Cambridge change, such that this change is not dependent on there ever existing any concrete objects or real changes. For example, it suffices that there be abstract objects that transiently exemplify properties of the form, being different from the obtaining state of affairs consisting of *the presentness of the instant t*. Abstract objects could also have such transient properties as occupying a present interval t_0, such that t_0 would have been occupied by one revolution of the earth around the sun and would have included the event of Jesus' birth, had these concreta existed. Abstract objects could also successively exemplify properties such as (*a*) occupying the present interval t_1 that is immediately later than the aforementioned interval t_0 and of the same length as t_0, (*b*) occupying the present interval t_2 that is immediately later than t_1 and of the same duration as t_1, etc.

These counterfactuals indicate that it is not logically or metaphysically necessary that concrete processes occur in order for time to have a metric. Consider the proposition expressed by 'there never occur concrete or real changes, but the abstract object O is enduring for a period of time that would be taken up by one revolution of the earth around the sun, if there were such an earth and sun'. This proposition is logically false or false of metaphysical necessity only if the concept expressed by 'enduring for a period of time' is equivalent (logically or metaphysically) to the concept expressed by 'enduring for a period of concrete or real changes'. However, we have seen in the preceding part of this chapter that these two concepts are not equivalent.

Given that the *possibility* of there being time before the big bang can be derived from our preceding arguments, the important question to address in this section is whether it is *probable* or *improbable* that there is time before the big bang.

We can begin by generalizing from fundamental a posteriori principles about the uniformity and regularity of nature, which have appeared in the sciences over the centuries in such forms as the law of mass-energy conservation and Newton's first law of motion (that something x continues to remain at rest, or continues to move in a straight line, unless there is something y different from x that acts upon x and causes it to discontinue its state of rest or straight motion). Such fundamental a posteriori principles imply that reality is not arbitrary, in a relevant sense of 'arbitrary'.

One relevant sense is embodied in the following general principle: if some object x exists at some temporal interval t of a finite length L, and (a) there is nothing in x's nature that makes it necessary or probable that x does not exist in an immediately prior temporal interval t' of length L, and (b) there is nothing in x's nature that makes it necessary or probable that x does not exist in an immediately subsequent interval t'' of length L, and (c) there is nothing y different from x upon which x is dependent, such that y would make it necessary or probable that x did not exist at the earlier time t' (and the same for the later time t''), then (d) x probably exists at t' and t''.

This may be called the *principle of non-arbitrary duration*. This principle is a contingent and a posteriori truth, since there are many possible worlds in which all or most things arbitrarily begin to exist or cease to exist, i.e. begin or cease to exist for no internal or external reason of *any* sort (not just causal or sufficient reasons).

The principle of non-arbitrary duration implies a corresponding a posteriori and contingent principle about time itself. For any interval t of any finite length L, there probably are intervals of length L both immediately before and immediately after t if there is nothing internal to time's nature or external to time that necessarily or probably precludes this extended existence.

There is nothing internal to time's nature that precludes this extended exist-ence (i.e. theories of the necessary finitude of time such as Craig's can be shown to be unsound[7]). The only apparent external reasons for precluding this extended existence (before the big bang and after the big crunch) rest on the false verificationist philosophy that belongs to GTR and orthodox QM (and some of the more recent quantum gravity cosmologies). For example, one verificationist reason is that observable light clocks (or what would appear to be a light clock to an observer) cannot exist at the big bang or big crunch (or during the first and last intervals of 10^{-43} second). But once these veri-ficationist reasons are rejected, and the concreta-independent nature of time is recognized, there is no case for there being reasons external to time (reasons involving the behaviour or existence of concrete or abstract items that exist in time) that necessarily or probably prevent time's existence before the big bang and after the big crunch.[8]

If the above reasoning is sound, we are entitled to believe what may be called *the principle of discrete temporal infinitude*. This principle states that:

> For any temporal interval t of any finite length L, there
> probably are intervals of the same length as t both
> immediately before t and immediately after t.   (P1)

For example, for each hour t, there probably is an hour t' immediately pre-ceding t and an hour t'' immediately succeeding t. This principle is useful inasmuch as it enables us to generalize about time's extension beyond the physical processes accessible to observation.

The a posteriori principle (P1) implies that the past and future are prob-ably infinite, i.e. that past time and future time each contain at least $\aleph_0$ (aleph-zero) hours. Principle (P1) is satisfied if (but not only if) time is probably infinite and has the 'standard' order type of infinite time, omega-star plus omega, $\omega^* + \omega$. This is the order type of the positive and negative integers in their 'natural' order $\{\ldots -3 \quad -2 \quad -1 \quad 0 \quad 1 \quad 2 \ldots\}$. The number zero cor-responds to the present hour, the negative numbers correspond to past hours, and the positive numbers to future hours. However, we shall see there are reasons to think that infinite time probably does *not* have this order type.

---

[7] Craig's theory is discussed in Craig (1979), Smith (1987), Smith (1989), Craig (1991), Craig and Smith (1993), Smith (1993b), and Craig (1993).

[8] Could time be causally dependent on the creative activity of a supernatural concrete object, God? My arguments in this chapter that abstract objects exist in time apply also to God; if God exists, God exists in time. Brian Leftow (1991) has given a number of arguments for the thesis that if God exists in time, God cannot create or destroy time. I have also argued in Smith (1996) that it is logically impossible for God to cause anything. Given these theses, we need not take into account divine activity (if there is any divine activity) when considering whether there are external reasons for time beginning or ending.

The principle of non-arbitrary duration, applied to time, is properly generalized to intervals of any length, finite or infinite. (As I use the phrase 'infinite interval', a temporal interval is infinite if and only if it contains at least an aleph-zero number of equal-lengthed finite temporal intervals (e.g. an aleph-zero number of hours).) If it is possible for there to be infinite temporal intervals before and after any infinite interval t, there probably are infinite intervals before and after t if there is nothing external to time or internal to time's nature that necessarily or probably prevents infinite intervals from existing before and after t. If there are no preventative factors of these sorts, we have a second principle about time:

> If it is possible for there to be an infinite temporal interval
> prior (and subsequent) to any infinite interval t, then there
> probably is an infinite interval prior (and subsequent) to t. (P2)

Let us call this *the principle of temporal unboundedness*. This principle is true if there are no internal or external factors that necessarily or probably prevent time from being infinitely extended beyond any given infinite interval. The same considerations that apply to the unboundedness of finite intervals and (P1) also apply here, entitling us to believe (P2). That is, GTR, QM, and quantum gravity theories give no valid and non-verificationist reason to reject (P2). Moreover, time is not dependent on the existence of concrete objects or real changes. And there is nothing internal to time's nature that makes it probable or necessary that infinite time is 'bounded' in the sense relevant to (P2).

The principle of temporal unboundedness implies that infinite time probably does not have the order type $\omega^*+\omega$, since it is possible for there to be infinite intervals earlier (and later) than an interval with this order type. It is possible for a temporal sequence to have the order type $\omega^*+\omega+\omega^*+\omega$, which is the order type of the following series:

$$\ldots -4 \quad -2 \quad 2 \quad 4 \ldots \qquad \ldots -3 \quad -1 \quad 1 \quad 3 \ldots$$

There exists a set of numbers with this order type, and there is no logical or metaphysical difficulty with supposing that the set of all hours is isomorphic to this set of numbers.[9] The hour -2 immediately succeeds hour -4, the hour 2 immediately succeeds hour -2, hour 4 immediately succeeds hour 2, but an infinite number of subsequent hours elapse before hour -3 becomes present. Hour -3 is infinitely later than hour 4 and is immediately earlier than hour -1, which is succeeded by an infinite number of hours $1, 3, 5, 7, \ldots$ The principle of discrete temporal infinitude is satisfied, since for each hour

[9] This is argued in some of the works mentioned in n. 7, especially in Smith (1989, 1993*b*).

t, there is an hour immediately earlier than *t* and another hour immediately later than *t*.

The most frequently stated objection to the claim that time could have this order type is that the hour −3 could 'never' be reached from the hour 4. This statement is true if 'never' means not after any *finite* amount of time has elapsed. But it is false if 'never' means not after an *infinite* amount of time has elapsed. I think the real objection these philosophers have in mind is that it is 'counter-intuitive' to think that time could have this order type. But Cantor has shown us that there are 'counter-intuitive' truths about transfinite cardinals, ordinals, and other order types. In the present case, no relevant difference between numbers and times has ever been established, i.e. a difference that would show times *cannot* correspond to these numbers.

Given the possibility that time could have the order type $\omega^*+\omega+\omega^*+\omega$, if time probably begins with the temporal series $\{\ldots-3 \quad -1 \quad 1 \quad 3\ldots\}$, without any reason to make this beginning probable or necessary, then the principle of non-arbitrariness (applied to infinite time in (P2)) would be false. If (P2) is true, then there probably is an infinite interval before this series. This condition is met if the series $\{\ldots-3 \quad -1 \quad 1 \quad 3\ldots\}$ is preceded by the temporal series $\{\ldots-4 \quad -2 \quad 2 \quad 4\ldots\}$. For similar reasons, the principle of temporal unboundedness is violated if there probably is no temporal series earlier than $\{\ldots-4 \quad -2 \quad 2 \quad 4\ldots\}$, since it is possible for there to be an earlier series with the same order type, $\omega^*+\omega$. This line of reasoning proceeds infinitely, leading to the hypothesis that there are an infinite number of temporal series of order type omega-star plus omega, both before the series $\{\ldots-3 \quad -1 \quad 1 \quad 3\ldots\}$ and after this series.

Why should we believe there probably is an infinite sequence of infinite temporal series? Principle (P2) is a version of a principle of non-arbitrary duration, generalized to apply to infinite intervals of time. If the 'local' infinite interval *t* in which we live has the order type $\omega^*+\omega$, and there is nothing internal or external to time that necessarily or probably prevents an infinite temporal interval from preceding the infinite interval *t*, and yet there is no such preceding interval, then this is an arbitrary fact. The sequence of infinite temporal intervals would *arbitrarily begin* with the maximal infinite subsequence of *t* that has the order type ω^* $\{\ldots-3 \quad -1\}$ and would *arbitrarily end* with the maximal infinite subsequence of *t* that has the order type ω $\{1 \quad 3\ldots\}$. There is no difference between a sequence of finite equal-lengthed intervals (e.g. hours) beginning or ending arbitrarily at some interval and a sequence of infinite intervals beginning or ending arbitrarily at some interval, *apart from the finite/infinite difference*. But this difference is not relevant to considerations of whether or not time begins or ends arbitrarily, for 'time arbitrarily begins with a first finite interval *t*' and 'time arbitrarily begins with a first infinite interval t' both imply 'time arbitrarily begins with a first interval t'.

According to *habitual* philosophical and scientific patterns of thinking and talking, 'time has the order type $\omega^*+\omega$' does not imply 'time begins', let alone 'time arbitrarily begins'. But habitual patterns of thinking and talking need not be true patterns. If time has the order type $\omega^*+\omega$, then time does not 'begin', if 'begin' means has a first interval of some given finite length (e.g. a first hour). However, if we are not talking merely about finite intervals, but about all intervals, finite and infinite, then time does 'begin' in the sense that it has a first infinite interval, namely, the maximal interval with the order type ω^*. If there is nothing in the nature of time, and no external reason, that requires or makes it probable that time does not extend beyond a given interval (finite or infinite), then time's extension is arbitrarily interrupted.

Principles of non-arbitrariness override Occam's razor ('do not multiply entities beyond necessity'). There is no necessity to postulate times later than noon, 22 August 1999. Reality would contain fewer entities if we supposed that time ended then, rather than lasted longer or had an infinite future. However, reality would be arbitrary if time ended at noon, 22 August 1999, and this arbitrariness justifies the belief that time will not end at this date. The same applies to the infinite case.

It may be objected that there is a relevant difference between the finite and infinite cases. If time stopped at noon, 22 August 1999, this would be arbitrary since the current boundary conditions of the universe (the current amount and arrangement of mass-energy etc.), in conjunction with the laws of nature, imply that it is probable that the universe will continue to exist after this time. The arbitrariness of time ceasing to exist would be equivalent to laws of nature arbitrarily ceasing to hold at this time and the causal processes in nature inexplicably ceasing to exist. By contrast, there are no (known) concrete causal processes connecting infinite intervals that would be arbitrarily interrupted, and no (known) laws of nature that would be arbitrarily violated if there is only one interval with the order type $\omega^*+\omega$.

However, this objection presupposes the verificationist metaphysical theories of STR, GTR, and QM that I have been arguing against in this chapter. These metaphysical theories that imply time's beginning to exist, ceasing to exist, or continuing to exist are logically, metaphysically, or nomologically dependent on concrete objects and events and upon the causal processes connecting concrete events. (Even the most 'anti-verificationism' inclined interpreters of STR and GTR, such as Nerlich, regard time as depending on the existence of space.) I have argued that the existence of time is not dependent on concrete events or causal processes or anything concrete at all, including space. Accordingly, the arbitrariness of time beginning or ending cannot be defined in terms of a concrete causal order arbitrarily beginning or ending or of a natural law governing physical events (or space) being arbitrarily suspended; rather, it needs to be defined in terms of a *concreta-independent* time, as I have done in (P1) and (P2). Given the concreta-independent nature of time, the principle of non-arbitrary duration is justifiably applied to finite temporal intervals in (P1) and to infinite temporal intervals in (P2). The seeming 'intuitive implausibility' of (P2) is based on traditional habits of thinking about infinite time and the assumption that time is concreta-dependent. If time is concreta-independent, as I have argued, we need to change our habitual patterns of thinking about infinite time.

Let us give the name alpha, α, to the order type of any set isomorphic to the infinite set with the order type of $\{\ldots \omega^*+\omega+\omega^*+\omega \ldots\}$. Is this order type special? Of course, infinite mathematical sequences of mathematical sequences, each of which is infinite, need not have the order type alpha. But temporal series with the order type of many of the variously ordered transfinite mathematical sequences do not conform to the principles of discrete temporal infinitude and temporal unboundedness. Unlike the sequence with the order type alpha, most infinite sequences of infinite sequences violate the principle of discrete temporal infinitude. For example, a sequence with the order type $\{\ldots \omega+\omega+\omega+\omega \ldots\}$ violates the principle of discrete temporal infinitude, since each subsequence with order type omega has a first interval of each finite length L (e.g. a first hour), and thus has a finite interval of length L that is not immediately preceded by another interval of that length. In fact, the principle of discrete temporal infinitude implies that time's order type cannot be any ordinal number (recall that not all order types are ordinals). The conjunction of (P1) and (P2) narrowly restricts the possible order types that infinite time can possess.

Does infinite time probably have the order type alpha? Note that all the infinite intervals in the beginningless and endless sequence $\{\ldots \omega^*+\omega+\omega^*+\omega \ldots\}$ of infinite intervals compose a single infinite temporal interval T, which has the order type alpha. It is possible for there to be another interval T' with the order type alpha that is earlier than T, consistently with the principle of discrete temporal infinitude. Given the principle (P2) of temporal unboundedness, this indicates that there probably is such an interval T'. The same holds for T', such that there are an infinite number of intervals with the order type alpha that are earlier than T (and an infinite number of intervals with the order type alpha that are later than T).

The infinite temporal interval $T1$ composed of $\{\ldots T'+T+T'' \ldots\}$ has an order type which we may name alpha-one. By analogous reasoning, we conclude that it is probably the case that $T1$ is preceded by an earlier temporal interval $T1'$, with order type alpha-one, and that $T1$ is succeeded by a later temporal interval $T1''$ of order type alpha-one. There are an infinite number of temporal intervals with the order type alpha-one and the interval $T2$ composed of all these intervals has the order type alpha-two. This line of reasoning proceeds ad infinitum.

Principles (P1) and (P2) are jointly satisfied if time has the structure described in the preceding paragraphs. I shall not here address the question of whether it is possible for (P1) and (P2) to be satisfied by a time with a different structure.[10]

One infinite series of order type $\omega^*+\omega$ is temporally connected to another series by earlier/later relations. Are there causal connections? The causal relations are between two infinite series, not between any member of one series and any member of another series. There is a partial analogy to causal relations among intervals of continuum-many instantaneous events. One instantaneous event E is not caused by any event that occupies the immediately preceding instant, since there is no immediately preceding instant. Nor is there any instantaneous event E' that causes E; if E' caused E, the causal relation between E' and E would 'bypass' an infinite number of subsequent instantaneous events that separate E' from E. But this 'bypass' is impossible. Consequently no instantaneous event can bear a causal relation to another instantaneous event. Rather, only a temporally extended process, which occupies an interval of continuum-many instants, can cause another process, which occupies another interval of continuum-many instants. Causal relations obtain between infinite series of instants, not between two instants.

There is a partial analogy here to our infinite series of temporal sequences, even though they are not densely ordered. Causal relations do not obtain between finitely long events or processes in one series of the order type $\omega^*+\omega$ and

---

[10] Another question is whether there is any justification for believing that there are more than aleph-zero temporal sequences with the order type $\omega^*+\omega$. Consider two successive units of time, where a 'unit' means a sequence with the order type $\omega^*+\omega$. We may name one of these units '1' and the later unit '2'. It is possible for there to be a unit prior to unit 2 and later than unit 1, such that the three units are isomorphic to the three numbers 1, 1.5, 2. Perhaps, for *any* two successive units t and t', it is probable that there is a unit after t and before t', such that the units of time are densely ordered or isomorphic to rational numbers.

Further, it may be that the units of time are isomorphic to a set of real numbers, such that the sequence of all units of time is continuous and not merely dense. According to Robinson's (1966) theory of non-standard real numbers, between any two standard real numbers n and n', there are an infinite number of non-standard real numbers that are greater than n and less than n'. It may be that the entire sequence of the units of time is isomorphic to a sequence of standard and non-standard reals that is open at both ends.

One problem with this suggestion is that the justificatory basis for the principle of temporal unboundedness (P2) is the principle of non-arbitrary duration, which is about discretely ordered intervals, i.e. intervals with *immediate* predecessors and successors. Thus, further argumentation would be needed to show that there are more than aleph-zero units of time.

finite processes in a later series with this order type. Rather, they obtain between two infinitely long physical processes. The following scenario is consistent with the currently available observational evidence about the universe. Suppose we have a contraction of all galaxies and other matter from infinity, the contracting process having the order type ω^*, that ends in a bounce at time zero ('the big bang'), with another expansion of matter to infinity, with the expansion having the order type ω. The contraction–expansion process has the order type $\omega^*+\omega$:

$$\ldots -5 \quad -3 \quad -1 \quad 0 \quad 2 \quad 4 \quad 6 \ldots$$

The 'bounce' or 'big bang' occurs at the small finite interval denoted by '0'.

This entire contraction–expansion process, composing an interval with the cardinal number aleph-zero, causes another contraction–expansion process that composes a later interval with the same order type and cardinal number. This next interval is also composed of an infinitely long contraction followed by an infinitely long expansion phase.

We must avoid the temptation to think that the series $2 \quad 4 \quad 6 \ldots$ eventually reaches a state that is 'the maximal infinite expansion', a state named aleph-zero, and that the universe then begins contracting from this state of maximal expansion. Rather, there is no maximal state of expansion; for every state of expansion, there is a larger state of expansion. No state of expansion stands in a causal relation to any state of contraction in the next series, $\ldots -6 \quad -4 \quad -2 \quad 0$. Rather a process consisting of an infinitely contracting and then infinitely expanding matter causes the next interval to be composed of an infinitely contracting and then infinitely expanding matter. If the arguments in this chapter are sound, and we adopt the principle of symmetry (if all else is equal, symmetry gives us a universe that is infinitely contracting and expanding in each subsequence of order type $\omega^*+\omega$), then the present empirical evidence about the universe is consistent with the suggestion that the infinite sequence of infinite temporal series is a sequence of contraction–expansion processes.

There may be a consistency between this theory and the observational evidence, but the principle of symmetry is not sufficient by itself to enable us to draw conclusions about what probably is the case.

If time has an infinitely infinite structure, then a new and difficult subject of inquiry is opened, namely, what sorts of concreta (*if any*) occupy the infinitely many infinite temporal intervals that are earlier or later than our 'local' interval. A more fundamental question is: what sort of theoretical criteria should be used to determine what sorts of concreta (if any) occupy some or all of the infinite intervals other than our local interval, and how much relative weight should be placed on each criterion?

The principle of symmetry (taken by itself) suggests that every infinite interval is symmetrical in its concrete nature. This corresponds to the theory just explained. If our local temporal series has the order type ω*+ω, and consists of a universe that contracts for an infinite amount of time (corresponding to the {... −2   −1} sequence), 'bounces' at time 0, and then expands for an infinite amount of time (corresponding to the {1   2 ...} sequence), then earlier and later temporal series also have the order type ω*+ω and contain similar contracting and expanding universes.

But *the principle of plenitude* (taken by itself) suggests that every possible concretum, or every possible kind of concretum, exists in at least one of these infinite intervals (given the restriction that the plenum is internally consistent).

By contrast, *Occam's razor* (taken by itself) suggests that all the infinite sequences but our local infinite sequence contain only necessarily existent abstract objects.

But other principles, such as explanatory power, simplicity, and the like, are also relevant. Thus, it is far from obvious what is 'going on' in the infinite sequences that are probably earlier or later than our local infinite sequence.

9. Summary

In summary, I have argued in this chapter that (*a*) the theory of time in orthodox quantum mechanics, the Special Theory of Relativity, and the General Theory of Relativity are based on a false verificationist metaphysics, and that the philosophical parts of these theories should consist of a realist metaphysics, that (*b*) observable physical clock processes do not measure the metric and topology of time, that (*c*) all abstract objects exist in time, that (*d*) the simultaneity and presentness of physical events are not relative to a reference frame, but absolute, that (*e*) past and future time are infinite, and that (*f*) time consists of an infinite sequence of infinitely long temporal intervals.

REFERENCES

Callender, Craig, and Weingard, Robert (1994), 'The Bohmian Model of Quantum Cosmology', in David Hull et al. (eds.), *Proceedings of the Philosophy of Science Association*, 1: 218–27.

Craig, William Lane (1979), *The Kalam Cosmological Argument* (London: Macmillan).

—— (1991), 'Time and Infinity', *International Philosophical Quarterly*, 31: 387–410.

Craig, William Lane (1993), 'Smith on the Finitude of the Past', *International Philosophical Quarterly*, 33: 225–31.

—— (forthcoming), *God, Time and Eternity*.

—— and Smith, Quentin (1993), *Theism, Atheism and Big Bang Cosmology* (Oxford: Clarendon Press).

Einstein, Albert (1952), 'The Foundations of the General Theory of Relativity', in A. Einstein et al. (eds.), *The Principle of Relativity* (New York: Dover Publishers): 111–64.

—— (1961), *Relativity: The Special and General Theory* (New York: Crown Publishers).

Forbes, Graeme (1985), *The Metaphysics of Modality* (Oxford: Clarendon Press).

Isham, Christopher (1988), 'Creation of the Universe as a Quantum Tunneling Process', in R. J. Russell et al. (eds.), *Physics, Philosophy and Theology* (Vatican City: Vatican Observatory Publishers): 375–408.

Leftow, Brian (1991), *Time and Eternity* (Ithaca, NY: Cornell University Press).

Nerlich, Graham (1994), *The Shape of Space* (Cambridge: Cambridge University Press).

Owens, David (1992), *Causes and Coincidences* (Cambridge: Cambridge University Press).

Penrose, Roger (1989), *The Emperor's New Mind* (Oxford: Clarendon Press).

Plantinga, Alvin (1974), *The Nature of Necessity* (New York: Oxford University Press).

Quine, W. V. O. (1974), *The Roots of Reference* (Cambridge, Mass.: Open Court Publishers).

Robinson, Abraham (1966), *Non-standard Analysis* (Amsterdam: North-Holland Publishers Co.).

Schlegel, Richard (1980), *Superposition and Interaction* (Chicago: University of Chicago Press).

Sidelle, Alan (1995), review of Quentin Smith's *Language and Time*, in *Review of Metaphysics*, 48: 679–81.

Smith, Quentin (1987), 'Infinity and the Past', *Philosophy of Science*, 54: 63–75.

—— (1989), 'A New Typology of Temporal and Atemporal Permanence', *Noûs*, 23: 307–30.

—— (1990), 'Time and Propositions', *Philosophia*, 20: 279–94.

—— (1993*a*), *Language and Time* (New York: Oxford University Press).

—— (1993*b*), 'Reply to Craig: The Possible Infinitude of the Past', *International Philosophical Quarterly*, 33: 109–15.

—— (1996), 'Causation and the Logical Impossibility of a Divine Cause', *Philosophical Topics*, 24: 169–91.

—— (1997*a*), *Ethical and Religious Thought in Analytic Philosophy of Language* (New Haven: Yale University Press).

—— (1997*b*), 'The Ontological Interpretation of the Wave Function of the Universe', *Monist*, 80: 160–85.

Swinburne, Richard (1981), *Space and Time* (New York: St Martin's Press).

—— (1996), review of Quentin Smith's *Language and Time*, in *Philosophy and Phenomenological Research*, 56: 486–9.

Tooley, Michael (1987), *Causation: A Realist Approach* (Oxford: Clarendon Press).

Vallicella, William F. (1995), 'No Time for Propositions', *Philosophia*, 25: 473–80.

Zahar, Eli (1983), 'Absoluteness and Conspiracy', in R. Swinburne (ed.), *Space, Time and Causality* (Dordrecht: D. Reidel): 37–41.

—— (1989), *Einstein's Revolution* (Cambridge: Cambridge University Press).

8

Freedom and the New Theory of Time

L. NATHAN OAKLANDER

Although McTaggart's famous article on 'The Unreality of Time' initiated the contemporary debate between tensed and tenseless theories of time, it was not until the last two or three decades that the literature and interest in the issue blossomed. Undoubtedly, one of the reasons why the tenser–detenser debate in the philosophy of time has recently attracted so much attention is its connection with other important issues in philosophy. The area I wish to connect with the tenser–detenser controversy is metaphysics or, more specifically, the problem of human freedom. Indeed, the connection between time and freedom is so close that many have argued that in order to preserve human freedom we must abandon the tenseless theory of time and adopt the tensed theory. The arguments for this position are not new. Perhaps the first argument connecting time and freedom is implicit in Aristotle's discussion of the sea battle, and it is still common for defenders of the tensed theory to argue that the tenseless view is incompatible with human free will.[1]

The criticism that the tenseless theory leads to a denial of human freedom has been levied on several different fronts. Some have argued that the tenseless theory implies a *logical* threat to freedom.[2] Since detensers maintain that all events exist determinately at the moment they do, regardless of what moment it is, they are committed to the determinacy of truth-value, and so to the principle of bivalence. But then, the logical fatalist argues that if, for any statement S and any time t, S is either true at t or false at t, it follows that every statement, including statements about the future, is either *now* true or *now* false and that, therefore, the future (like the past) is fixed and unalterable.

I wish to thank Ron Hoy, Ned Markosian, Robin Le Poidevin, Hugh Mellor, and Alan White for their very helpful comments on earlier versions of this chapter. I have also benefited from discussions of this chapter at the Mike Morden Memorial Colloquium at Oakland University (in Michigan) and at West Virginia University.

The research for this chapter was supported (in part) by a fellowship from the faculty development fund of the University of Michigan-Flint.

[1] For recent examples see Lucas (1989), Shanks (1994), and Yourgrau (1992).

[2] See, for example, Steven Cahn (1982). For a good discussion of the connection between time, truth, and fatalism see Bernstein (1992).

According to others, the tenseless theory implies a *metaphysical* threat to freedom. For if all objects in time have a tenseless existence, then nothing comes into existence or ceases to exist, and autonomous control over what does or does not exist is an illusion.[3] Moreover, if detensers maintain, as arguably they must, that objects persist by *perduring*, that is, persisting things are wholes composed of temporal parts, then nothing *really* changes.[4] Without real change, however, we cannot bring about a change in the properties of an object, and human creativity and freedom is lost.

Furthermore, some have claimed that the tenseless theory implies a *causal* threat to our freedom. Since, on the tenseless view, all events exist at the time and place they do, with the qualities they have, they are determinate in their existence and statements about them are determinate in their truth-value. If, however, all temporal objects, and the statements about them, are *determinate*, then everything that takes place is *determined* to be exactly the way it is. Conversely, a critic of tenseless time may argue that the only reason for thinking of the future as determinate is the belief in determinism. But if all events, including human actions, are determined, then they are unfree.[5]

Tensers also argue that the tenseless theory implies a *phenomenological* threat to freedom since it cannot account for our experience of freedom.[6] We experience the future as being an open realm of possibilities, but for the detenser all events are equally real, there being, as Yourgrau puts it, '*a symmetry of past, present and future with respect to facticity*' (1992: 46). However, if the future is as real as the present, and so already exists or already is a fact, then how can the detenser account for our experience of the role we play in creating the (*not yet existing*) future?

Finally, there is what I will call the *argument from science*. This argument incorporates virtually all of the objections to tenseless time contained in the preceding arguments.[7] It runs something like this. The mathematical representation of physical objects in contemporary physics entails a tenseless theory of time. On this representation time is a space-like dimension along which physical objects lie. It is then argued, as Lucas once put it, that 'an inevitable concomitant of this approach [is] that we take a block view of the universe in which the future course of events is already laid out as a path in Minkowski

---

[3] Cf. Shanks (1994).

[4] Carter and Hestevold (1994) argue that the tenseless theory implies the doctrine of temporal parts. For arguments that purport to show that nothing really changes on a temporal parts ontology see Lombard (1986, 1994) and Geach (1968). For a reply to Lombard see Heller (1992). For a reply to Geach see Oaklander (1984). For a fully developed causal account of tenseless change within a temporal parts ontology see Le Poidevin (1991).

[5] Łukasiewicz (1967) seems to argue in this fashion. For critiques of the inference from determinate to determined see Grünbaum (1976), Oaklander (1984), Smith and Oaklander (1995), and Williams (1951).

[6] See Lucas (1989: 8–9).   [7] Cf. Yourgrau (1992).

space-time',[8] which in turn implies the logical, metaphysical, and phenomenological threats to human freedom previously delineated.

Clearly, any complete defence of the tenseless theory of time must address the various charges that it is incompatible with human freedom. To consider all the issues I raised, with the detail and complexity they deserve, would require much more space than I have for this chapter. For that reason, my aim is more modest and the limitations of this essay should be clear at the outset. I am not attempting to give a complete account of human freedom, but only to show that the nature of tenseless time, when properly understood, does not entail a denial of human freedom and that only a host of misinterpretations could lead one to believe that it did.

1. *The new theory of time*

The theory that I wish to defend has been called the 'static', 'stasis', and 'spacelike' theory of time, or, less pejoratively, the 'tenseless' or 'B-theory' of time. I prefer to call it the 'new theory' of time. The other labels, and especially terms such as 'spacelike' and 'tenseless', can easily lead to misinterpretations. For example, the notion of time being 'tenseless' is problematic because it is ambiguous. Tenselessness is a feature of language, and (for some) it is also a feature of reality. Indeed, the distinctive feature of the new theory of time is that tensed (or tenseless) language and TENSED (or TENSEless) reality must be clearly separated, although proponents and opponents of the tenseless theory have not always kept them separate. Early advocates of the tenseless theory argued that metaphysical TENSE was eliminable from reality because grammatical tense was translatable without loss of meaning from ordinary language. On the new theory, the existence of grammatical tense as represented in ordinary language and thought is not to be confused with ontological TENSE, as represented in a tensed theory of time. Thus, on the new theory, the need for tensed discourse in ordinary language does not imply the existence of tensed properties in reality. And conversely, the rejection of ontological TENSE, that is, the denial of tensed properties and tensed facts, does not imply the translatability of tensed language in ordinary discourse. In other words, given that we are representing reality, a tensed language is eliminable in terms of a tenseless one, even though a tensed language cannot be translated in terms of a tenseless one.[9]

---

[8] Lucas (1986: 130). Lucas himself does not accept this implication of modern physics and adopts a version of the open future theory of time which he argues is compatible with physics. For a recent discussion of the open future theory see Markosian (1995).

[9] For recent articles on the new theory and criticism, see the essays by Mellor, MacBeath, Oaklander, Williams, and Smith in part i of Oaklander and Smith (1994).

To see why tense and tenselessness as applied to language and time must be kept distinct, consider the following sentence-types:

> Oaklander is (present tense) working on a paper on time
> and freedom, (A)

and

> Oaklander is (tenselessly) working on a paper on time
> and freedom on 13 October 1996. (B)

According to some tensed theorists, (A) literally changes its truth-value in that it expresses a proposition which has different truth-values at different times. For defenders of the new theory, on the other hand, the difference between (A) and (B) consists in two related facts. First, (B) is time indexed and (A) is not, and because of that fact, different tokens of (A) can have different truth-values, whereas different tokens of (B) each have the same truth-value. More generally, sentences with a tenseless copula have an 'unchanging' or 'permanent' truth-value; they are 'always' true (if they ever are), whereas sentences with a tensed copula can 'change' their truth-value (i.e. can have tokens with different truth-values) depending on when they are tokened. But why is that so? How is the tenselessness of the copula to be interpreted?

There are three likely possibilities. On one interpretation the tenseless copula is literally without tense, so that (B) is read as:

> Oaklander's working on a paper on time and freedom has the
> property of presentness on 13 October 1996. (B′)

Alternatively, we can interpret the tenseless copula as *omnitensed* so that (B) is read as:

> Oaklander is (was and will be) working on a paper on time
> and freedom on 13 October 1996.[10] (B″)

Or, we can interpret the tenseless copula as *timeless* as in 'Red is a colour' or 'Two plus two is equal to four' so that (B) is read as:

> Oaklander is (timelessly) working on a paper on time and
> freedom on 13 October 1996. (B‴)

Any of these models provide a basis for the 'permanent' or 'unchanging' truth-value of (B), and the 'fact' that (B) is 'always' true, but they can also easily give rise to trouble.

Clearly, (B′) is a tenseless sentence in the sense specified. Different tokens of (B′) always have the same truth-value. But either (B′) means that

---

[10] Padgett (1992) interprets the tenseless copula as omnitensed, and Smith (1993) interprets it as conjunctively (present and future) tensed. According to the new theory, however, a tenseless copula lacks any tense.

E is simultaneous with 13 October 1996, or it means that E occurs on 13 October 1996 and E has *presentness*. The former interpretation is innocuous whereas the latter commits one to the reality of TENSE and for that reason is unacceptable to the detenser.

Similarly, if one interprets the tenselessness of the copula in (B) as omnitensed (B″), and if one then confuses omni*tensed* language with temporal reality, a tenseless truth like (B) will imply the existence of tensed properties; a thesis adamantly denied by new theorists. Furthermore, if one confuses the 'tenseless' sentence (B″) with its truth-conditions (or the fact in virtue of which it is true), then it is a short step to the conclusion that the basis in reality for the permanent truth of (B″) is a state of affairs that *always exists* or *exists at every moment*. Thus, if (B″) represents the nature of time and not just the language we use to communicate about it, then (B″) implies that I am now (always have been and always will be) working on this paper on 13 October 1996, and that is absurd!

The third interpretation of the tenselessness of the copula is equally troublesome. For if the tenseless copula in (B) is construed as the *timeless* copula of, say, 'Two plus two is equal to four', and tenseless language is confused with timeless reality, then the objects in the B-series would be timeless, and time and change would be unreal.

C. D. Broad, who thought often and deeply about the problem of time and change, seems to have misinterpreted tenseless time in the ways I have just explained. In his *Examination of McTaggart's Philosophy* Broad says:

The [tenseless] theory seems to presuppose that all events, past, present, and future, in some sense 'co-exist,' and stand to each other *timelessly* or *sempiternally* in determinate relations of temporal precedence. But how are we to think of this 'co-existence' of events? It seems to me that the events and their temporal relations are thought of either by analogy with timeless abstract objects such as the integers in their order of magnitude, or by analogy with *simultaneous persistent particulars*, like points of a line in spatial order from left to right. Neither of these analogies will bear thinking out. (1938: ii. 307; my emphasis)

I agree that neither of these analogies will bear thinking out, but I disagree that the new theory of time is committed to either of them. On the new theory, temporal relations between events are in some sense analogous to the relations that obtain between universals, but we need not claim that the terms of temporal relations are to be thought of as universals, that is, as timeless abstract objects. Like relations among universals, temporal relations between and among events, and the facts that they enter into, are not located at any time or in any place. Yet, it does not follow that the *terms* of temporal relations coexist timelessly in the way in which universals do. Nor does the eternal truth of 'E_1 is earlier than E_2' require E_1 and E_2 to be located at all times. Why, then, might Broad think that events are persistent particulars or timeless objects?

One possible explanation for Broad's mistake is that he does not clearly separate the language which states the truth-conditions of 'E_1 is earlier than E_2' from the reality to which that sentence corresponds. By construing the tenselessness of language as either omnitensed or timeless, and then confusing the language of time with the reality of time, one might easily be led to the conclusion that the terms of temporal relations are simultaneous persistent particulars or timeless abstract objects.[11]

A more recent example of the confusion between tenseless language and temporal reality occurs in the writing of Alan Padgett. Padgett argues that what he calls the 'stasis' theory is guilty of committing the fallacy of 'confusing the logical with the physical' (1992: 118). He claims that detensers confuse the logical with the physical when they argue that

since something is a fact at time T3, and it is thus always a fact and can be expressed by a true statement that is always true, then in some way the fact of 'the fact at T3' must always exist. (Padgett 1992: 119)

But what exactly is the difference between logical facts and physical facts, and do new theorists actually confuse these two notions? As we shall see, what Padgett says concerning this distinction reflects the blurring of language and reality to which I have been alluding.

Padgett writes,

There is a distinction between a 'fact' from a logical point of view and a 'physical' fact. A 'fact' from a logical point of view is the truth expressed by a true statement. Thus it is a 'fact' that 2+2=4, or that 'London is south of Cambridge.' (1992: 118)

In this passage, Padgett's notion of a logical fact equivocates on the phrase 'the *truth expressed* by a true statement'. On the one hand, the truth expressed is the fact or state of affairs that makes a statement true, for example the fact that 2+2=4. On the other hand, it means the sentence or statement which is true, for example 'London is south of Cambridge.'[12]

---

[11] Admittedly, in order to represent reality we must use language, and for that reason, temporal language and temporal reality are not entirely distinct. Nevertheless, the new theorist maintains that the language used to represent the metaphysical nature of temporal reality must be without tense since the reality of TENSE (as either properties or facts) entails unacceptable dialectical difficulties (i.e. McTaggart's paradox). For a discussion of McTaggart's paradox and criticisms of recent attempts to avoid it, see Oaklander and Smith (1994: part ii) and Oaklander (1996).

[12] In correspondence Padgett has replied that 'the fact that "London is south of Cambridge" is a *physical fact*; on p. 118 I am NOT talking about the sentence, "London is south of Cambridge" but rather the *physical facticity* that this sentence denotes' (e-mail message, 5 June 1996; my emphasis). However, if that is Padgett's view then he is confusing logical and physical facts, since in the passage quoted in the text, Padgett uses 'London is south of Cambridge' as an example of a *logical fact*.

It seems clear, however, that, for Padgett, a logical fact or 'truth' expressed by a true statement is not the fact to which it corresponds since he distinguishes logical from physical facts by maintaining that

A 'physical' fact . . . is an event or some state of affairs *in the world*. I will call such things 'physical states of affairs.' . . . Examples of a physical state of affairs would then be: my having the thought I am having *now*, the hotness of the sun, and the liquidity of the water in my cup. Now physical states of affairs can only be real or existing events. Thus the difference between the stasis and the process views of time can be put this way: is it *now* a physical state of affairs that the sun rises (tenselessly!) on 4 July 1776? The process theorist says 'No,' and the stasis theorist says 'Yes.' For the stasis view insists that, in some way or other, the event of the sun's rising exists (tenselessly) on 4 July 1776. (1992: 118; my emphasis)

Padgett interprets the new theorist to be claiming that since 'the sun's rising (tenselessly) on 4 July 1776' is a tenselessly true statement, it follows there now exists (always did exist and always will exist) the physical fact of *the sun's rising on 4 July 1776*. For that reason, he accuses the new theorist of confusing the logical fact or 'truth' which is always true, with the physical fact that exists only on 4 July 1776.

Admittedly, it is indeed a fallacy to argue that, simply because '*E* occurs at t_1,' *always* expresses a truth, it follows that it is now (always was and always will be) a physical fact that *E occurs at t_1*, but the new theorist does not commit it. Padgett could only think that it did follow by first confusing tenseless language with omnitensed language and then confusing omnitensed language with temporal reality. To see what other difficulties the failure to keep language and time separate can give rise to, I shall turn to the logical threat to freedom allegedly posed by the new theory, namely, fatalism.

2. *Logical fatalism*

Logical fatalism is based upon the principle of bivalence according to which for any sentence *S*, including those about the future, and any time *t*, *S* is either true at *t* or false at *t*.[13] If, however, at time *t*, a sentence about the future is

[13] An anonymous reader correctly noted that *the* principle of bivalence concerns truth *simpliciter* and *not* truth at a time. Nevertheless, one might confuse the two concepts of truth: because all tokens of a tenseless sentence are true (or false) whenever they are uttered or written, one might mistakenly infer that a tenseless sentence is *always* true, i.e. true at every time, and thus fallaciously conflate the principle of bivalence (for truth *simpliciter*) with the corresponding principle of bivalence (for truth at a time). We shall see that those who attribute logical fatalism to the new theory of time are guilty of that mistaken inference and subsequent confusion. Alternatively, if one holds a tensed view of time one might reject the concept of truth *simpliciter* in favour of the concept of truth at a time. Since, however, according to the

now true, or *already* true, then the future is *fixed* and there is nothing we can do to prevent it. And if, at time *t*, a sentence about the future is *now* false, or *already* false, then there is nothing we can do to bring it about. In either case, as Aristotle puts it, 'nothing is or takes place fortuitously, either in the present or in the future, and there are no real alternatives; everything takes place of necessity and is fixed' (1941: 46).

The crucial assumption in this argument for fatalism is that if a sentence is now true at time *t*, then there now exists (or 'already exists') at time *t* a state of affairs in virtue of which it is true. Fortunately, the new theorist need not accept that assumption. Consider the future-tense sentence 'There will be a sea fight.' According to the new theory as I understand it, the fact in virtue of which that sentence is true will vary depending on the time at which it is uttered or inscribed. For example, if the sentence in question is asserted today, at t_1, then the state of affairs which exists if the sentence is true is *a sea fight occurs later than* t_1. However, that state of affairs is not located at the time at which the future-tense sentence is uttered, and it is not located at any later (or earlier) time. Of course something is located at t_1, namely, the inscription or utterance 'There will be a sea fight', and assuming the utterance is true, something is located at a time later than t_1, namely, the sea fight, but the whole state of affairs that *a sea fight occurs later than* t_1 is not located at t_1, is not located at t_2, and is not located at any earlier or later time. Indeed, it is not located *in* time at all.[14]

States of affairs which contain temporal relations between events are *eternal* in the sense of existing outside the network of temporal relations, but

---

detenser, all truth is truth *simpliciter*, any argument against the new theory which is based on the concept of truth at a time (as logical fatalism is) is question-begging and unsound. Recently, Michael Tooley (1997) has argued that an adequate conception of temporal reality requires both conceptions of truth *simpliciter* and truth at a time. For a criticism of Tooley's theory see Oaklander (forthcoming).

[14] Similarly, the fact that *Socrates is taller than Phaedo* exists outside time, but depends for its existence on Socrates and Phaedo, who exist in time. Robin Le Poidevin has suggested to me another argument for the claim that facts do not exist in time. Suppose that the fact (call it *F*) that *e occurs at t* exists at *t*. Then there is another fact, namely the fact that *F exists at t*, and so on *ad infinitum*. On the other hand, D. H. Mellor has objected to my saying of some facts that they exist outside time. He would say of some facts, not that they do not exist in time, but that they have no limited location in time, and likewise for space. Thus, the fact, call it *P*, that Mitterrand dies, is located in space and time: Mitterrand dies in Paris in January 1996. On the other hand, the fact, call it *Q*, that Mitterrand dies in Paris in January 1996 is not located in any region *R* of spacetime such that '*Q* in *R*' is true and '*Q* outside *R*' false. For Mellor, this is just a consequence of the fact that all of *P*'s temporal and spatial aspects are already stated in *Q*. As far as responding to the fatalist's argument against the new theory, Mellor's point may be well taken. My preference for treating temporal relational facts as outside time stems from the belief that temporal relations are simple, primitive existents, and atemporal universals. A defence of these views is, however, outside the scope of this essay.

not in the sense of existing (or persisting) throughout all of time. Thus, an 'eternal' entity is one that neither occupies moments of time, nor exemplifies temporal relations, nor has monadic temporal properties inhering in it. Rather, an eternal entity is related to time in the following way: it is a whole which contains successive parts. We could say that although an eternal entity is not contained in time, time is contained in it. Thus, the fact that *World War II is later than World War I* is eternal because, although not itself located in time (or a term of a temporal relation), it contains time (a temporal relation) as a constituent. Consequently, the truth of a future-tense sentence does not imply that the future 'pre-exists' in the present, or that the future 'already exists'.

Of course, if one confuses the tenseless sentence 'A sea fight occurs later than t_1,', which states the truth-conditions of 'There will be a sea fight' (uttered at time t_1), with the (TENSEless temporal relational fact) to which it corresponds, namely, *a sea fight occurs later than* t_1, then fatalistic worries will emerge. For the sentence describing the (TENSEless) truth-conditions of a future-tense sentence is tenselessly true, i.e. true *simpliciter*. However, if one confuses true *simpliciter* with true *at every time*, then one will conclude that the tenseless sentence is *always* true. If one then confuses the tenseless sentence with the state of affairs to which it corresponds, one might conclude that it is *always* existing. If, however, *a sea fight's occurring later than* t_1, or *a sea fight's occurring at* t_2, always exists and thus 'already' exists at t_1, that is, at a time before the sea fight occurs, then my choice is an illusion, for I can neither bring about nor prevent at t_2 what already exists at t_1. On the other hand, if one recognizes that a future-tense sentence is true in virtue of an event's occurring *later* than the time of the sentence, and not in virtue of anything that is located *at* the time of utterance, the difficulty vanishes.

My new-theorist response to logical fatalism rests upon the notion of an 'eternal' state of affairs. Alan Padgett has criticized my notion of 'eternal' temporal relations for again confusing the logical fact that '$S1$ is earlier than $S2$' which is always true with the physical fact that $S1$ and $S2$ always exist. Thus, he says:

The process theorist can agree, if need be, that 'facts' like $S1$-earlier-than $S2$ or Alan born-before-Carl are 'eternal' in that they are omnitensely true. Tensed sentence tokens affirming their existence all make statements which are true, always were, and always will be. From this it does not follow, however, that the episodes referred to by $S1$, $S2$ and $S3$ exist eternally, any more than it follows that Carl and I are 'eternally' born, or exist 'eternally.' Oaklander and others who follow this argument have committed the fallacy of confusing the logical with the physical. (1992: 120)

I agree that such an inference would be fallacious but deny that I make it. To say that '$S1$-is-before-$S2$' is always true does not imply that $S1$ and $S2$

are either eternal, sempiternal, or timeless objects, and Padgett could only think that the new theory would imply that it did by falling prey to the confusion between temporal language and temporal reality discussed throughout this essay.

Although Padgett's argument does not undermine the existence of 'eternal' facts (or timeless temporal relations), one might still object that the existence of events that are TENSElessly located at later moments takes away our freedom. For if there is a fact that, say, *I vote for Gore in 2000* (or that *I vote for Gore later than t_1*) then, arguably, that fact *necessitates* the occurrence of my voting for Gore in the year 2000, or synonymously, that it is not within my power, in 1996, to prevent my voting for Gore in 2000. It is not at all clear, however, that the existence of an eternal fact necessitates anything. The necessity does not follow from the principle of bivalence. The law necessitates that one of a pair of contradictory statements is true, but it does not imply that either '*P*' is necessarily true or 'not-*P*' is necessarily true. If I vote for Gore in 2000, then my present statement (in 1996) 'I will vote for Gore in 2000' is true, and if I do not vote for him then my present statement 'I will not vote for Gore in 2000' is true. Thus, which statement is true depends on whether or not the fact *I vote for Gore (approximately four years) later than 15 November 1996* exists or does not exist, but whether or not that fact exists depends upon what I will choose to do on election day, November 2000. In other words, it is my later decision that determines which of two contradictory statements about the future is true, since it is my later decision that determines what eternal fact exists. Thus, the existence of eternal facts is not incompatible with our having it within our power to bring about or prevent certain events. Indeed, it is because we do have it within our power (because we do choose or do not choose) to bring about or prevent certain events that certain eternal facts exist.[15]

One might still wonder how the existence of facts that are not located in time can depend on something that is located in time. How can the 'eternal' fact that *I vote for Gore later than 15 November 1996* depend for its existence on something that occurs in November 2000? Although eternal facts

---

[15] Perhaps one will ask how, say, Jones is *able* to not eat dinner at t_1, if it is true (or a fact) that Jones does eat dinner at t_1. Very briefly, I would say, that the notion of 'ability' or what we 'can' do is ambiguous. Relative to one set of facts someone may be able to do something which relative to another set of facts one is unable to do. Thus, given the fact *Jones eats dinner at t_1*, there cannot also be the fact that *Jones does not eat dinner at t_1*, but given certain other facts which we ordinarily take to be relevant to what we can or cannot do, it *can* be the case that Jones does not eat dinner at time$_1$. Given facts about Jones's physical capacity to drive a car and to stop by the local pub, his propensity to have a drink, and the typical rush-hour traffic around dinner time, and so on, he *can* avoid eating dinner at time$_1$, but he will not. See Lewis (1993).

do not exist in time (i.e. as terms of temporal relations), they do contain temporal relations and their relata as constituents. Thus, the fact that I am going
to vote for Gore in four years *includes* my decision next year to vote for
Gore. That decision is an event that exists in time, and if it does not occur
on 5 November 2000, then there is no such fact as my voting for Gore in
2000. For that reason, the truth of a future-tense sentence does not entail the
existence of a fact which in turn determines my choice. Rather, my choice
in November 2000 determines what (TENSEless) fact exists, and hence what
future-tense sentence is true.

Interestingly, Storrs McCall, a tenser, who once argued that propositions
about the future must have an indeterminate truth-value, has recently changed
his mind. And what he says on this topic closely reflects the new theorist's
view:

We shall say that the truth of an empirical proposition supervenes upon events in the
sense of being wholly dependent upon them, while at the same time events in no
way *supervene upon* truth. Thus the truth of the proposition that X is in Warsaw
town square at noon next Friday depends upon what happens next Friday, and in this
way the sting of 'logical determinism' is drawn. . . . What is true today depends upon
what happens tomorrow, not the other way round. The set of true propositions in no
way determines what the future is like. Instead, what the future is like determines
the set of true propositions. (1994: 14)

3. *Metaphysical fatalism*

While the logic of temporal language does not commit the new theorist to
denying human freedom, the metaphysical implications of the new theory may.
In a recent and provocative paper Niall Shanks (1994) claims that there are
important connections between the free will/determinist debate and some issues
in the metaphysics of time. Specifically, he argues that friends of libertarian
incompatibilism can find no support in modern indeterministic physics
because (1) modern physics implies the S-theory (presumably for spacelike
or static theory) of time and (2) if the S-theory is true, then there is no human
autonomy.[16] In this section I will consider Shanks's main arguments in
defence of (2)—the thesis that the S-theory implies the denial of human
autonomy—and argue that they suffer from a pernicious spatialization of
(TENSEless) time, and a confusion of common sense with ontology.

---

[16] That modern physics implies the tenseless theory is a highly contentious point that
Shanks assumes without argument. Recent discussions of the issue occur in Smith (1993), Stein
(1991), and Tooley (1997).

To see what is involved in these claims and to justify them let us turn to Shanks's argument in support of (2). Shanks believes that in order to have genuine human autonomy the universe must be open in three senses of the term: one causal and two ontological. In the causal sense, the universe is open$_1$ if indeterminism is the case, that is, if the location and configurations of physical objects at some time t are underdetermined by (or not all lawfully connected with) earlier states of the universe. In an ontological sense, the universe is open$_2$ if not only is it indeterministic (open$_1$), but also the locations and configurations of physical objects are *autonomously changeable*. 'That is, [location and configuration] must be properties of a physical object which admit of alteration and adjustment as a consequence of intervention by, or interaction with an autonomous being' (1994: 48–9). In other words, in an open$_2$ universe, the actions of a causally efficacious (underdetermined) agent are capable of bringing about a *real* change in a physical object. In a third sense, the universe is open$_3$ if there is *ontological indeterminacy*, meaning that 'there are many possible futures compatible with their present circumstances, and it is by no means determined or "fixed" at present which of these will come to pass' (1994: 55) and thus that the objects of history *do not* tenselessly exist at their respective moments with their respective properties, i.e. the new theory is false. Autonomous action requires openness in each of these senses. For if the universe is either causally *determined* (closed$_1$) or ontologically *determinate* (closed$_3$) then it is closed$_2$ since no autonomous being is capable of 'bringing about' or 'causing' an existential or qualitative change in a physical object, and genuine human autonomy is impossible.

Shanks claims that even though the S-theory *does* allow for the possibility that the universe is indeterministic (or open$_1$), it renders the future fixed (or closed$_3$), and human autonomy impotent (closed$_2$). For on the S-theory, 'time is literally a space-like dimension along which an object is extended' (1994: 58), and since 'the distinct spatial parts of an object at some given time t enjoy the same existential status . . . the distinct temporal parts of an object will likewise be on a par, ontologically speaking' (1994: 52). However, if the four-dimensional shapes of physical objects are nothing more than 'coexisting temporal parts' (1994: 56), tenselessly existing in different regions of spacetime, 'then the objects of history neither begin to exist nor cease to exist' (1994: 55). Unfortunately, if objects of history neither begin nor cease to exist, but exist *simpliciter*, then what the future will be is determined or fixed at present, and the belief in autonomous changeability and human autonomy is an illusion.

Shanks's argument, familiar as it is seemingly persuasive, is nevertheless rooted in a web of confusions. We can begin to see what they are by noting that Shanks believes our commonsense intuitions and ordinary language talk about time have implications for ontology. As he says,

In order to see that the information expressed by the tenses has ontological consequences, it suffices to consider the following fragments of ordinary language: 'Elizabeth I *existed* four hundred years ago', 'Elizabeth II *exists*', and 'Elizabeth III *will* (possibly) *exist*'. According to common sense, of the three monarchs, only Elizabeth II actually exists. The tenses thus enable ordinary speakers to express existential differences between the objects of history. (1994: 51)

Since, however, the tenseless theory treats all objects as *ontologically on a par, tenselessly existing* at the times they do, Shanks concludes that the S-theory must deny our intuitions that only the present exists and that there are existential differences between the objects of history.

It seems to me, however, that Shanks's line of reasoning confuses grammatical tense as represented in ordinary language and thought, with ontological TENSE as represented in the new theory of time. That is, Shanks assumes that, since we use *grammatical* tenses to indicate existential differences between objects of history, it follows that it must be the tensed properties of *pastness*, *presentness*, and *futurity*, or the tensed facts, It was the case, It will be the case, or It is the case that *X is present*, that distinguish objects that exist from those which do not. In other words, on the tensed theory, the commonsense platitude 'only the present exists' carries ontological weight. On the new theory, on the other hand, there are no such properties (and no such facts), and *in that respect*, time *is* analogous to space. Just as there is no monadic property of *hereness* that distinguishes this place, or the place where I am located, from any other place, there is no monadic property of *presentness* that distinguishes this time, or the time when I am located, from any other time. Thus, time is TENSEless, and all objects are ontologically on a par in that there are no tensed properties that distinguish those events that exist from those that do not. It does not follow that tensed language and thought is translatable in terms of tenseless language, or that we could communicate without tensed language. Nor does it follow that our intuitions about time are mistaken (although they need to be properly interpreted) or that tensed language is false. Indeed, once our ordinary language of time is distinguished from a metaphysically perspicuous language of time, it can be shown how and why our temporal intuitions are true.

As I now write (at 4.00 p.m., 15 November 1996), only those events which are (roughly) simultaneous with my utterances exist and are located at that date. Events which will occur (or TENSElessly occur) on 23 November 1996, eight days later than my writing, do not yet exist in that they are not located at (and do not TENSElessly exist on) 15 November, and events which did occur (or TENSElessly occur) on 14 November 1996, one day earlier than my writing, no longer exist in that they are not located at (and do not TENSElessly exist on) 15 November either. Thus, on the new theory our ordinary beliefs that the future is not yet and the past is no longer and the present is now are true but innocuous. Future events are not yet because they are located at times later than my speaking, past events are no longer because they are located at times earlier than my speaking, and only the present exists in the deflationary sense that only those events which are simultaneous with my speaking are located at 4.00 p.m., 15 November 1996. These are the tenseless truths that underlie the intuitions that there are existential differences between objects of history and that only the present exists; the existence of tensed properties or facts has nothing to do with it.

In addition to 'tenselessly existing', two other phrases that give rise to trouble are 'equally real' and 'coexisting'. There is, of course, a phenomenological difference between space and time: although we experience objects at different spatial locations as being 'equally real' (i.e. occurring at the same time), we do not experience objects at different temporal locations as being *equally real* or *coexisting*. This phenomenological claim is used by tensers to support, on the side of common sense, the claim that the present has a privileged position in the temporal spectrum (i.e. only the present exists), and on the side of ontology the claim that objects in time exist if and only if they exemplify *presentness*.

The fallacy in this line of reasoning consists in its confusing TENSEless time with space. We do experience objects elsewhere in space as coexisting equally with objects that are here, but we do not experience past (earlier) and future (later) events as coexisting with those events that are present. Thus, if one thinks that time is literally like space, then the claim that four-dimensional objects are composed of 'coexisting temporal parts' or that all objects in TENSEless time are 'equally real' will be taken to deny this phenomenological fact, and reduce temporally separate events to simultaneous parts of a presently existing whole. In that case, however, the universe is ontologically determinate (closed$_3$), since what the future will be is fixed at present, and human autonomy is lost.

Fortunately, such an illicit spatialization of time, along with its dire consequences for human autonomy, is not a necessary part of the new theory of time. On the tenseless theory as Russell (1915), (the early) C. D. Broad (1921), Clifford Williams (1994), Oaklander (1984, 1995, and 1996), and other detensers have conceived of it, temporal relations are primitive and unanalysable relations, and the difference between spatial and temporal relations is an irreducible qualitative difference. The experience of earlier and later upon which the experience of time and change is built is primitive, and is intrinsically different from our experience of one coloured patch being to the left of another differently coloured patch. For that reason, objects coexist in time in a fundamentally different sense from the way they coexist in space. Thus, time possesses a quality that space does not possess: spatially separated objects can exist at the same time, but temporally separated events cannot occur at the same time. It is this crucial difference between the terms of spatial and temporal relations that is the phenomenological basis of the sense of privilegedness that the present moment possesses.[17]

Of course, there are crucial similarities between space and time as well. Just as temporally separated objects cannot exist at the same time, spatially separated objects cannot exist at the same place, and just as spatially separated objects can exist at the same time, temporally separated objects can exist at the same place. Nevertheless, neither of these analogies between space and time should lead one to deny that on the new theory there is an intrinsic difference between time and space or assert that the terms of the temporal relations of earlier/later coexist in the way in which the terms of the spatial relations of left/right coexist.

Closely connected with the particular interpretation Shanks gives to the S-theory is the claim that 'if the S-theory is true, then the objects of history neither begin to exist, nor cease to exist' (1994: 55). Once again, if this is true, then nothing could be brought into existence and autonomous existential control (that is, the ability to bring or not bring something into existence) would be impossible. But it is not true.

Consider, for example, the experience of a headache ceasing to exist. It involves first having a headache and at the same time being conscious that one is having a headache. It involves second the consciousness of no longer having a headache. This involves both the awareness of my having various thoughts and feelings, and not having a headache. It also involves the memory of a headache that is not located now, at this moment, but is located (or exists TENSElessly) at an earlier moment. In other words, if I am aware at time$_1$ that I have a headache, and I am aware at a later time$_2$ that I do not have a headache, and I remember my headache existing at time$_1$, then I am having an experience of my headache ceasing to exist.

What this account of our experience of time makes clear is that the ceasing to exist of a headache (or any other event, for that matter) is a process that takes place at two moments: the last moment of its existence and the first moment of its non-existence. Thus, on the detenser's reading, a headache's ceasing to exist over the interval t_n–t_{n+1} is its being located up to t_n and thus

---

[17] Cf. Williams (1994: 365–6). For a causal analysis of tenseless temporal relations see Le Poidevin (1991) and Mellor (1981).

making the present-tense belief 'My headache exists (now)' true up to t_n, and false at t_{n+1} (and later). Similarly, a headache's beginning to exist at t is nothing more than its being located at t and not earlier, and thus making the present-tense belief 'My headache exists (now)' true at t and false earlier.

To this it might be objected that since the knowledge of my headache ceasing to exist requires that the *tensed beliefs* 'My headache exists (now)' and 'My headache did exist' are both true (at different times), there must be *tensed facts* to account for their truth. The inference, however, is fallacious. For if a belief or judgement is indexical, as it is if it is tensed, then its truth-conditions are token-reflexive. So all it takes to make a token of the tensed belief 'My headache exists (now)' true is that the headache occurs simultaneously with the belief. And all it takes to make a token of the tensed belief 'My headache did exist' true is that the headache ended before I had the belief (or that the belief is held after the end of the headache). If, however, we can make sense of objects coming into existence and ceasing to exist on the new theory, existential control (as reflected, for example, in determining 'whether or not to make an apple pie for supper' (1994: 57)) is still within the grasp of the new theorist.

Human autonomy also requires qualitative control and Shanks attempts to show that to be impossible on the four-dimensionalist account of change, since the four-dimensionalist must reject the commonsense intuition that objects persist with a numerical identity through time. For

> if the S-theory is true, then Elizabeth II, for example, will be said to have temporal parts in Paris *and distinct temporal parts* in London, it no longer makes sense to say that the object in Paris at t1 is one and the same object that is now in London at t2—instead there are distinct temporal parts of a four-dimensional whole at these space-time locations. (1994: 52)

To uncover the confusions that permeate this argument, note that his statement 'it no longer makes sense to say that the object in Paris at t1 is one and the same object that is now in London at t2' is ambiguous. For the phrase 'the object in Paris (*P*) at *t*1' may mean 'the temporal part *P*-at-*t*1' (or, more simply, '*p*1'). Or, it may also mean 'the entity *P*, of which *P*-at-*t*1 (*p*1) is a part'. Clearly, if we mean the former, then it is misleading, indeed false, to assert that *P*-at-*t*1 (*p*1) is the same person as *P*-at-*t*2 (*p*2), but this does not imply that the person of which these two different stages are temporal parts is not the same person at these different times. For on the second interpretation we can say that the person *P* which at one stage in its life (*p*1) is in Paris is the very same person *P* that at a later stage (*p*2) is in London. Thus, the existence of four-dimensional objects with temporal parts is compatible with thinking of these objects as being the subjects of qualitative change, and with our being able to bring about some of those changes.[18]

It seems, therefore, that some of the crucial intuitions that are employed in the argument for the incompatibility of the new theory of TENSEless time and human freedom, for example that only the present exists, that there are existential differences between the objects of history, that objects can begin to be and cease to be, and that things change, *can* be explained on the new theory. What, then, of Shanks's claim that

the modality which some people believe the future to have—in the sense that they believe there are many possible futures compatible with their present circumstances, and it is by no means determined or 'fixed' at present which of these will come to pass—must, when analyzed in S-theoretic terms, arise instead merely from ignorance. (1994: 54–5)

If by 'determined' *at present* Shanks means that, on the S-theory, the locations and configurations of physical objects at some future time *t* are causally determined (closed$_1$) by (or lawfully connected with) present states of the universe, then he is falling prey to the confusion between 'determined' and 'determinate' he sought to avoid. And if by 'fixed' *at present* Shanks means that, on the tenseless theory, there *now* exist facts that ground the truth of future contingents, then he is wrong. For, on the new theory, the truth of '*A* will do *X*' when uttered at time *t* corresponds (in part) to the fact that *X* exists (*TENSElessly*) *later than t* (and not to any fact that is located *at t*). On the other hand, if he means that the future is fixed like the past in that we cannot change it, then I would agree. But to say that we cannot change the future is not to say that we do not have a hand in making it, or that our choices do not bring it about.

4. The phenomenology of freedom

Nevertheless, there is an intuition that we all share according to which the past is unalterable, fixed, and already settled and thus no longer within our power to control, whereas we are free to make choices that will determine how the future will be. What account of the phenomenological asymmetry of openness with respect to the past and future can the new theorist give? In this concluding section I will mention three possibilities congenial to the new theory that are worthy of further exploration.

---

[18] Heller has argued that 'once we accept that there really is such a thing as the extended four dimensional whole, there is no good reason to deny that it, and not just its temporal parts, does have properties at various times' (1992: 700). Cf. Oaklander (1987–8).

The asymmetry of openness is analysed by David Lewis as an asymmetry of counterfactual dependence. The future depends counterfactually on the present in a way in which the past does not so depend. As Lewis puts it,

What we can do by way of 'changing the future' . . . is to bring it about that the future is the way it actually will be, rather than any of the other ways it would have been if we acted differently in the present. . . . Likewise, something we ordinarily *cannot* do by way of 'changing the past' is to bring it about that the past is the way it actually was, rather than some other way it would have been if we acted differently in the present. (1991: 53)

This suggestion by Lewis is quite plausible and it is compatible with both the new theory and with determinism. My major hesitation concerns Lewis's analysis of counterfactual dependence and the subsequent commitment to possible worlds that it entails. Thus, in order to avoid the thorny problem of giving an adequate account of the truth-conditions of counterfactuals, I therefore prefer, although I will not here explore in any detail, two suggestions put forth by Broad during his (early) detenser stage.

Broad says that there are two senses in which the past is fixed and unalterable, while the future depends, in part, on our volitions. He says:

(i.) However much I may know about the laws of nature, I cannot make probable inferences from the future to the past, because I am not directly acquainted with the future, but I can make probable inferences from the past to the future; i.e., although every possible proposition about the future is even now determinately true or false, I may be able to judge now, from my knowledge of the past and present and of the laws of nature, that some propositions about future events are much more likely to be true than others. . . . (ii.) I know with regard to certain classes of events that such events never occur unless preceded by a desire for their occurrence, and that such desires are generally followed by the occurrence of the corresponding events. But the existence of a desire for *x* does not increase the probability that *x has* happened. If it did we might be said to affect the past in exactly the same sense in which we can affect the future. Thus the assertion that we can affect the future but not the past seems to come down to this; (a) that propositions about the future can be inferred to be highly probable from a knowledge of the past and present, but not conversely, because of our lack of direct acquaintance with the future; and (b) that the general laws connecting a desire for *x* with the occurrence of *x* always contain *x* as a consequent and never as an antecedent. (1921: 335)

According to Broad's first point, the asymmetry of openness is epistemological. The second point seems to me to be more convincing. Our experience of the openness of the future is based upon the awareness that the desire for the occurrence of a certain event is something that is causally efficacious or lawfully connected with the desired event only when the event is *later* than the desire. And our experience of fixity stems from our knowledge that a desire or wish that something *had* happened is not lawfully connected with that event's having happened. For these reasons, we do not experience the past as something we have any control over, whereas we do experience the future as something over which we do have some control.

I think the new theorist's account of coming into existence and ceasing to exist is relevant here. For the openness of the future is, in part, based on our experience of an event that does not yet exist (i.e. is not located at the time of my decision to bring it about) coming into existence (or being located at a time) after I decide to bring it about. As I argued earlier, the experience of coming to be and ceasing to be is one that the new theorist can accommodate. On the other hand, the experience of the closed past or the fixity of the past stems from the experience or realization that my desire (at t_2) for a certain event to have happened in the past (at t_1) does not bring that event into existence. Similarly, my decision (at t_2) to end an unpleasant event (at t_3) may be lawfully connected with that event ceasing to exist (or not being located at t_3), but my wish (at t_3) for an unpleasant past event (at t_2) not to have happened (or not to be located at t_2) does not make a difference. These experiences, which form the basis of the asymmetry of openness between the future and the past, can be explained on the new theory.

Although much more could and should be said concerning the phenomenology of freedom, these brief remarks suggest the direction a new theorist of time could consistently take. In conclusion, I hope to have developed the new theory sufficiently to convince you that it is not the threat to human freedom that its detractors have claimed it to be.

REFERENCES

Aristotle (1941), *De Interpretatione: The Basic Works of Aristotle*, ed. Richard McKeon (New York: Random House).

Bernstein, M. H. (1992), *Fatalism* (Lincoln: University of Nebraska Press).

Broad, C. D. (1921), 'Time', in J. Hastings (ed.), *Encyclopedia of Religion and Ethics* (New York: Scribner Sons): 334–45.

—— (1938), *Examination of McTaggart's Philosophy*, vol. ii, pt. i (Cambridge: Cambridge University Press).

Cahn, S. (1982), *Fate, Logic and Time* (New Haven: Yale University Press, 1967; repr. Atascadero, Calif.: Ridgeway Publishing Company).

Carter, W., and Hestevold, S. (1994), 'On Temporal Passage and Temporal Persistence', *American Philosophical Quarterly*, 31/2: 269–83.

Geach, P. T. (1968), 'Some Problems about Time', in P. F. Strawson (ed.), *Studies in the Philosophy of Thought and Action* (Oxford: Oxford University Press): 175–91.

Grünbaum, A. (1976), 'The Exclusion of Becoming from the Physical World', in M. Capek (ed.), *The Concepts of Space and Time* (Boston: D. Reidel Publishing Co.): 471–500.

Heller, M. (1992), 'Things Change', *Philosophy and Phenomenological Research*: 695–704.

Le Poidevin, R. (1991), *Change, Cause and Contradiction* (London: Macmillan; New York: St Martin's Press).

Lewis, D. (1991), 'Counterfactual Dependence and Time's Arrow', in Frank Jackson (ed.), *Conditionals* (New York: Oxford University Press): 76–101.

—— (1993), 'The Paradoxes of Time Travel', in R. Le Poidevin and M. MacBeath (eds.), *The Philosophy of Time* (Oxford: Oxford University Press): 135–46.

Lombard, L. (1986), *Events: A Metaphysical Study* (New York: Routledge & Kegan Paul).

—— (1994), 'The Doctrine of Temporal Parts and the No Change Objection', *Philosophy and Phenomenological Research*, 54/2: 365–72.

Lucas, J. R. (1986), 'The Open Future', in Raymond Flood and Michael Lockwood (eds.), *The Nature of Time* (Oxford: Basil Blackwell): 125–34.

—— (1989), *The Future: An Essay on God, Temporality and Truth* (New York: Blackwell).

Łukasiewicz, J. (1967), 'On Determinism', in Storrs McCall (ed.), *Polish Logic* (Oxford: Oxford University Press): 19–39.

McCall, S. (1994), *A Model of the Universe* (Oxford: Clarendon Press).

Markosian, N. (1995), 'The Open Past', *Philosophical Studies*, 79: 95–105.

Mellor, H. (1981), *Real Time* (New York: Cambridge University Press).

Oaklander, L. N. (1984), *Temporal Relations and Temporal Becoming* (Lanham, Md.: University Press of America).

—— (1987–8), 'Persons and Responsibility: A Critique of Delmas Lewis', *Philosophy Research Archives*, 13: 181–7.

—— (1996), 'McTaggart's Paradox and Smith's Tensed Theory of Time', *Synthese*, 107: 205–21.

—— (forthcoming), 'Review of Michael Tooley's *Time, Tense and Causation*', *Mind*.

—— and Smith, Q. (1994) (eds.), *The New Theory of Time* (New Haven: Yale University Press).

Padgett, A. (1992), *God, Eternity and the Nature of Time* (New York: St Martin's Press).

Russell, B. (1915), 'On the Experience of Time', *Monist*, 25: 212–33.

Shanks, N. (1994), 'Time, Physics and Freedom', *Metaphilosophy*, 1: 45–59.

Smith, Q. (1993), *Language and Time* (New York: Oxford University Press).

—— and Oaklander, L. N. (1995), *Time, Change and Freedom* (New York: Routledge).

Stein, H. (1991), 'On Relativity Theory and the Openness of the Future', *Philosophy of Science*, 58: 147–67.

Tooley, M. (1997), *Time, Tense, and Causation* (Oxford: Clarendon Press).

Williams, C. (1994), 'The Phenomenology of B-Time', in Oaklander and Smith (1994: 360–72).

Williams, D. C. (1951), 'The Sea-Fight Tomorrow', in Paul Henle, H. M. Kallen, and S. K. Langer (eds.), *Structure, Meaning and Method: Essays in Honor of Henry M. Sheffer* (New York: The Liberal Arts Press): 282–306.

Yourgrau, P. (1992), *The Disappearance of Time* (New York: Cambridge University Press).

9

Morality, the Unborn, and the Open Future

PIERS BENN

The conflict between tensed and tenseless theories of time has ramifications in a variety of areas of philosophical concern. Moreover, as Thomas Nagel[1] has frequently stressed, the problems at the heart of the dispute about time and tense—one of which arguably, though not indisputably, concerns the very possibility of a wholly objective and perspective-free grasp of reality—are also at the heart of other philosophical issues, such as the mind–body problem, the problem of free will, and the nature of moral obligation. The question of whether statements containing indexicals can be given truth-conditions that avoid indexicals (even if there is a change of sense) is one of the problems at the core of many of these disputes. Many writers take the basic claim of tenseless theorists of time to be counter-intuitive and problematic, and devote themselves either to showing that the tenseless view is mistaken, or to showing how, despite the contrary appearances, certain important intuitions can be preserved in spite of the truth of the tenseless theory. My question, however, starts from the opposite angle. Could there be problems uniquely generated by the *tensed* theory of time—the view that states that ascription of tensed properties to dates reflects objective and irreducible features of reality? And if so, does the tenseless theory really avoid these problems, or related ones? One such set of problems concerns the asymmetry of concern for past and future individuals, and seems to generate ethical problems about the grounding of our concern for future individuals. If these problems are irresolvable, they serve to rebut some typical motivations (even if not grounds) for adopting the tensed theory. They compete for our attention with the various better-known difficulties associated with the tenseless view.

I shall not go into detail on the nature of the tensed/tenseless dispute and the main arguments canvassed for each position, as these are expounded elsewhere. But there is one feature of the debate which is worth bringing out at once. This is the dispute about the 'reality' of dates other than the present—especially the future. It is a hallmark of the tenseless conception of time that in it no ontological distinction is made between events or states of affairs on

---

[1] See especially Nagel (1986: ch. iv).

the basis of whether they are future, present, or past. Crudely put, the future is 'as real' as the present and the past, much as objects that are spatially on my left are no more or less real than objects that are on my right. In fact, this analogy sometimes causes the tenseless conception to be known as the spatialized conception of time. Thus, it is not the case, according to the tenseless view, that events pass from non-existence (in the future) to non-existence (in the past) via a fleeting (or duration-less?) moment of existence when they are 'present'. No changes of tense really occur at all, any more than London intrinsically passes from being 'here' to being 'not-here' when I take the train from London to Leeds.

Things may, however, be quite different on the tensed view. One version of this view does hold that the future, if not the past, is unreal; that it has no reality until it is present (or perhaps, also, past), in other words, until it ceases to be future. It is worth noting that this denial of the reality of the future is not strictly entailed by the tensed view: it is possible both that futurity is a genuine property of dates, and that propositions about future events have a determinate truth-value. Moreover, even according to the type of tensed view that denies the reality of the future, it is still the case (to put it with unavoidable looseness) that there will be a future, that events will happen, that some of them can be reliably predicted, and so on. But on this version of the tensed view, future contingent statements lack a truth-value. They are not determinately true or false, although *when* the putative states of affairs these statements refer to become present, these statements, recast in the present tense, will be *either* true *or* false.

With this tensed approach to the reality of the future in mind, one interesting question we face is this. In what ways can our moral obligations be 'forward looking', as opposed to backward looking? In particular, in what way can there be moral duties to future individuals? If we say that certain obligations are owed to future individuals, *to whom* exactly are they owed? Moreover, does it greatly matter whether we can give an answer to this, if we are to make sense of such obligations?

1. The unborn and the dead

The notion that we have moral obligations towards present individuals is of course unproblematic by comparison. However, there are familiar difficulties with the concept of obligations to past, now dead, people. Is there a good parallel to the problem of obligations to future persons, in the question of how there can be moral obligations to the dead? It must be admitted that some thinkers are troubled by the idea that there can be duties towards those who are no longer living. Those who reject the idea, or who try to recast the concept of duties to the dead in terms that avoid claiming that the dead literally have rights against the living, are often influenced by familiar Epicurean considerations. Such arguments from classical antiquity take different forms, but are centred on the thought that since the dead do not exist, no good or ill fortune can befall them. In particular, the dead cannot be wronged if promises made to them when they were alive are broken—for once dead, there is no one to whom such obligations are owed.

A familiar way of dealing with such arguments trades largely on the following conviction: that since many of the things we hope for and value are valued because we consider them valuable, quite independently of our awareness of them, or even our coexistence with them, it makes sense to hope and aim for certain things, even if we know that we shall not witness their actualization, and may even be dead when they come to be. I am not irrational if I hope that my property will fall into good hands after I die, and if I make a will to ensure this, it is because I consider that it will bring about something good. This well-known rejoinder to the Epicurean position is mainly intended to deal with the philosophical hedonism underlying Epicurus' view, which says simply that since the only intrinsically bad things for us are unpleasant experiences, and since death entails the complete absence of experience, death can therefore contain no evil for those who are dead. But such a response deals only with part of the problem, and not the most fundamental part. For an executor who carries out my instructions after my death does so on *my* behalf. That is to say, it is not merely because the instructions in my will happen to be good ones to follow that the executor enacts them. It is primarily because they are my instructions, and therefore to follow them is construed as a duty to me. It is this element in the consideration that really brings us to the nub of the difficulty identified by Epicurus. By all means (they may say) carry out the former wishes of people who are now dead, if those wishes are for worthwhile things—for after all, this may have been why people who are now dead wanted these things when they were alive. But do not construe this as any duty owed to them. To be the recipient of either justice or injustice, as one might be when one's reasonable wishes are followed or ignored respectively, one must—at the very least—exist. If the dead do not exist, then nothing can now befall them, and hence they cannot be the recipients of just or unjust actions committed now.

Though such arguments have a sophistical feel, there is remarkably little agreement among contemporary philosophers over how exactly they are to be refuted. For example, Roy Perrett (1987) thinks that the correct response is to affirm that interests can outlive their bearers, and that harms said to be done to the dead are really harms done to their interests, whilst Harry

Silverstein (1980) thinks that the solution to the problem lies in adopting a four-dimensional model of spacetime, allowing that the dead eternally exist, in virtue of spanning a particular segment of McTaggart's B-series. This diversity of solutions suggests that the simple Epicurean point is more intractable than it might have seemed at first consideration. On the other hand, the fact that most philosophers believe that the difficulty can be overcome somehow shows, at least, how far removed Epicurus' conclusion is from ordinary intuitions concerning the moral status of the dead. For it is generally agreed that there is both some sense in which death is an evil for those who die, and that there is some place for notions of moral requirements with respect to the dead, even if of a merely indirect kind.

One reason for trusting that Epicurus' argument can be answered stems from the fact that even if people cease to exist once they have died,[2] there is at least no problem about establishing reference to them. Requests to do certain things for the sake of the dead do not seem incoherent, as we might know perfectly well on whose behalf such requests are made. I suspect that the fact that the dead can be specifically referred to contains the beginning of an answer to Epicurus. Moreover, we should bear in mind that such things as reputation, and certain events in one's personal history such as being given posthumous awards, can clearly continue after one's death. Perhaps more important, and less often mentioned, is that the idea of loyalty surely has a proper role here. It seems right to remain loyal to one's dead friends and relatives, not necessarily because one thinks they exist, or even can be literally harmed, but because loyal acts (such as defending a dead person's reputation) are fundamentally *affirmative* in nature: they defend and affirm the worth or dignity of someone who is now dead. Furthermore, as we shall see, a loyal gesture or deed is always on behalf of some specific person, and does not vaguely affirm the dignity or value of humanity in general. Such loyalty to the dead is possible because determinate reference to them is also possible.

2. *The open future and its problems*

We are faced, however, with a far more intractable problem when considering the moral status of another class of non-existents, whose non-existence has either a different nature, or a different explanation, from the non-existence of the dead. One way of pointing to the difference is to ask whether the

---

[2] Interestingly this is quite controversial—not only because some people believe in life after death, but also because bodies continue to exist immediately after death, and if persons are bodies, it seems highly implausible to say that persons cease to exist at death, except perhaps in the sense that they cease to be persons—which is something different.

unborn[3] are determinate objects of reference, in the way that the dead quite clearly are. Although Winston Churchill is dead, and although it is arguable that nothing now can be predicated of him, at least in the present tense, he is clearly a proper object of reference. By contrast, if the future is indeterminate, as is held by some proponents of the tensed view, then future individuals are at most possibilities, who cannot be determinately referred to.[4] Thus, when prospective parents look forward to bringing their hoped-for offspring into the world, their thoughts can concern only *some child* they will have; they cannot be directed upon any particular, specific individual. This is not for merely epistemological reasons, since the phrase 'our future child' does not really function as a definite description. Even if we make the unquestionably reasonable prediction that new persons will come into being, nevertheless there are not, on a certain tensed view, any specific persons presently awaiting creation. Or, alternatively, even if there are specific, determinate *possibilities*, as may be allowed by some version of modal realism, there may still be no determinate truth about which of these specific possibilities will become actual. Such a combination of modal realism and non-realism about the future may be implausible, but it is not clearly impossible. Indeed, it is for reasons similar to this that one may be tempted to lament the fact not only that all actual people die, but also that the vast majority of possible individuals never come to exist.[5]

This point may be elaborated thus. Imagine that we take seriously the view that there simply are no determinate truth-values belonging to statements about future individuals, and that we are considering the question of what moral responsibilities there might be towards future generations. If there are not, in principle, any determinate objects of reference in the domain of future persons, there cannot now be any obligations towards specific future individuals. That is, since names like Jack or Jill do not pick out particular future persons, it is incoherent to suppose that there are now obligations *to Jack or Jill*, construed as putative unborn individuals. We now apparently face a tough choice. Either we can jettison altogether the belief that we have moral obligations concerning future persons, for the reason just given. Or we can retain the notion that we have such obligations, and try somehow to get

[3] Some may prefer to talk of the unconceived rather than the unborn, but I shall refer loosely to the unborn as individuals who do not yet exist.

[4] It may be argued that if causal determinism is true, then one could in theory, though of course not in practice, determinately refer to future individuals by tracing their causal connections to present ones. A theist who combined belief in tensed time with determinism could account for God's foreknowledge of future individuals in this way. However, apart from problems with determinism, it is still unclear that this picture would allow for reference to existent future beings whose mode of existence is symmetrical to that of the dead.

[5] See for example Palle Yourgrau (1987).

round the difficulty. The first option is plainly counter-intuitive, for whatever qualifications we may have to introduce to the concept of such obligations, it appears outrageous to claim that the thought of future beneficiaries or victims of our present actions cannot enter into our moral deliberations at all. For example, to place a sophisticated time bomb in a major city, primed to go off a century later, when it would probably kill and maim scores of people, would surely be monstrous. Wishing to acknowledge this, we might take the second option. In that case, we are back with our problem of say-ing to whom our moral concern is due.

The credibility of the second option is reinforced in the light of how I, presently alive, have benefited or suffered from deeds done by my ancestors. It was a good thing that a particular ancestor did not squander all his money, since had he done so he would have made my life worse than it is. I can surely credit this ancestor with a sense of his duty to future generations, which led him to do things that now benefit me. This thought is achieved simply by parity of reasoning with the case of the future. However, there is the ques-tion of to whom the ancestor's duties were owed. Initially it seems obvious that they were owed to me, among others. But there is still a problem. For when the ancestor was alive, there was no sense at all in which I existed—not even as a possible object of reference. True, I came to exist later on, but on the 'open future' hypothesis we are entertaining, there was no fact about my future existence, before I actually began to exist. So, again, to whom did my ancestor owe this duty?

Imagine, now, that the advocate of the open future argues as follows. The ancestor owed obligations, in fact, not to me specifically, but to a range of *possible* individuals, which then included me. How we limit the range of these possibilities need not be decided here; perhaps there is a finite range, with limits dictated by the number of possible genetic permutations at the time just before my conception, or perhaps the limits are set by metaphysical considerations, or perhaps there are no limits at all. The point is that we have established a kind of reference for statements about the persons to whom the obligations were due.

But this seems to lead to another, fundamental problem with the doctrine of a genuinely open future. Even conceding that moral duties can be owed to possible future persons, it still seems that no particular privileges attach to those possible persons who do, in the event, become actual. The duties are owed equally to all the possibilities. It is immaterial that I, in the event, was born, and that various other possible persons were not. My ancestor had no more obligations towards me than he had towards any other possible (for him, future) person. And, as we can now make clear, it is this which might be said to be one of the major difficulties with a tensed, open-future conception of time.

But what, now, of the merits of the tenseless theory of time? If it is true, does it not provide a way of understanding how there are, after all, specific future individuals? The idea certainly looks promising, since on this view, the events ahead of us, which will contain people who do not yet exist at the time of writing, are no less real than events which are simultaneous with or earlier than the time of writing. Moreover, this supposition yields the interesting moral possibility that there are duties that we now owe to specific future individuals. Various alleged ecological misdemeanours, such as depleting the earth's resources and causing harmful pollution, not to mention other wrongs such as permitting the decline of education and culture, may now be seen as harming specific future persons, which may give us a cogent reason for taking such duties seriously which would otherwise be lacking. A tenseless theorist can happily admit that we cannot know the future as we can know the past, and that there is no causal chain of reference allowing us actually to refer to as-yet-unborn persons. But this may be put down to the direction of causation. It does not take away from the fact that not-yet-born persons are every bit as real as dead and living ones.

3. Digression: tenseless time and fatalism

Indeed, the importance of causation in both tensed and tenseless conceptions of time needs to be emphasized, in order to rescue the tenseless conception from an endorsement of fatalism, which, if valid, would arguably undermine moral obligations altogether, let alone obligations to future beings. Fatalism, sometimes known as logical determinism, is an obvious temptation for those attracted to tenseless time. This is because of its invitation to picture both future and past as equally determinate, equally real, and therefore equally unchangeable. Its appeal is to the commonsense conviction that the past is unalterable, because past events are 'fixed', 'determinate', or 'have already happened'. If the future can be shown to be equally determinate, such that there is similarly a set of truths concerning which logically contingent things will happen, then the impression is generated that whatever will happen, must happen, and is therefore in some way beyond our genuine control. Nobody would seriously dream of trying to make the past other than it was; why then, asks the fatalist, is it any more rational to try to make the future other than the way it was always going to be? And indeed, the view of time suggested to us by the tenseless conception, which invites us to conceive of reality as it would appear to an Augustinian God who is himself outside time, does make fatalism attractive. For such a being, the whole of reality is an eternal given, with the parts that we now call future every bit as determinate as the parts we call past, and there certainly does seem to be something arbitrary about the idea that we have more power to shape the former than the latter.

However, if fatalism were really forced upon us, then we would need to revise our view of how there can be moral responsibilities towards any future beings—or indeed, to any beings at all. Whilst there is room for disagreement among fatalists concerning the sustainability of counterfactual conditionals with respect to the events which actually happen,[6] fatalists are agreed that whatever happens, happens inevitably; nothing in past, present, or future can ever be other than it is. But it is hard to see how we can be genuinely responsible for what, in the fatalist sense, could not have been otherwise, and it is equally difficult to make sense of moral obligation if we cannot apply the concept of moral responsibility.

Fortunately, there is good reason to deny that the tenseless theory of time entails fatalism. The fact (if that is what it is) that future contingent statements possess determinate truth-values does not entail that those truth-values could not have been other than they are; indeed, according to a standard (and surely sound) response, it is up to us to determine the truth-value of some future contingent propositions right up to the moment when the events referred to occur. On this account, if I shall choose to eat cake rather than bread for breakfast tomorrow, then the statement 'PB will eat cake on 24 August 1995' was always true; but it is still up to me to make it the case that this proposition has always been false. Resistance to this thought comes partly from a sense that the ability to determine the truth-value of a future-tensed statement, which was made in the past, would involve a kind of backward causation; or even worse, that it would entail the ability to *alter* the statement's truth-value—to make it other than it had once been. This is mistaken. If, ten years ago, someone uttered 'PB will eat cake on 24 August 1995', then he spoke truly if I do eat cake on that day, and falsely otherwise. But this does not prevent it from being up to me to decide, *after* he made the statement, whether or not the utterance *was* true. To put it briefly and familiarly: future-tensed statements may turn out to have been true when

---

[6] We may, perhaps, make a distinction between a fatalism which allows a role for counter-factual-sustaining causality and a version of the doctrine which does not. The first kind allows that events would have turned out differently if different causal antecedents had obtained, but denies that those antecedents could have obtained. Thus, on this understanding, statements like 'whatever he had done, he could not have avoided death on that day' are rejected. The second, rarer and perhaps (even) less coherent version, maintains *either* that causation between events has no real role in explanation, since everything is predetermined by a mysterious fate, *or* that there are certain specific events, typically death, which must occur as and when they do, even if we are free to do all we can to try to prevent them. In other words, any number of different attempts to avert fate are perfectly possible, but the attempts we actually do make are mysteriously bound to fail.

they were made, but it remains in our power, after they are made, to render them false.

We may seem to be labouring an obvious point, although there are philosophers who still deny it. But the insistence on the role of causation even within a tenseless view of time, and on our ability to shape the future even though it is ontologically on a par with the present and past, is crucial if we are to make anything of the idea that the existence of determinate future persons has moral implications for our present behaviour. Moreover, there is another important implication for our moral obligations towards future persons. This is that if the tenseless doctrine of time is relevant to the issue in the first place, such obligations can only be to those who actually will exist. It seems incoherent to condemn, on grounds of directly violating the interests of future persons, failure to create new people, or causing conditions which reduce the average number of children born to future generations compared to earlier ones, or even causing the human race to become extinct. Whilst there may well be different kinds of moral reasons for opposing actions with these effects, nevertheless no one who (if the tenseless view of time is true) was never going to exist can be wronged by not being created. If the tenseless view of time has any relevance to the moral standing of future persons, its relevance must be for those who actually will exist, or to put it differently, who eternally occupy points in time after the date at which I write these words. But given natural causal assumptions, the fact that these people will exist entails certain causes of their existence, and among those causes is a complex web of earlier human actions. Thus, moral considerations generated by the reality of future persons can apply only on condition that a vast range of actions by present and future people will be performed. If, then, anything important turns on the supposition that there are determinate facts about future individuals (rather than on the mere likelihood that there will be some, as yet non-specific, future persons), there is still almost nothing we can know about these future people. And the further in the future they are, the more difficult the problem becomes. The causal network is simply too complex for any useful predictions to be made.[7]

I suggest, then, that the moral implications of tenseless time must be, at best, more modest than we might at first suppose. We cannot actually pick out the individuals who will exist in the future, and we must recognize that the coming into being of any future person is always the outcome of a large

---

[7] In mentioning the complexity of the causal network I do not wish to imply that the tenseless view of time implies any particular view of causality, or any thesis of causal determinism. In fact, tenseless time is logically compatible with indeterminism. The truth-values of future contingents could derive merely from the reality of the future, rather than from any thesis of determinism.

set of coincidences, each of which is statistically extremely improbable. This means that the set of persons who actually will come into existence is a tiny subset of the set of possible individuals.[8] Thus, if the reality of future persons has any importance for morality, it seems only to be this: that it allows that talk of duties towards future persons is coherent, since the beneficiaries of these duties really do exist. In other words, it provides an answer to a problem symmetrical to that which worried the Epicureans, when they discussed the evil of death.

4. Asymmetries and their inevitability

But apart from grounding the reality of future individuals, does the tenseless view have any other advantage? There are further reasons for doubting that its implications are at all significant, as the following considerations suggest.

To begin with it is important to notice, after making all necessary concessions to the tenseless conception of time, that its view of ultimate reality should not be taken as an attempt to deny the importance of tense in our thinking about time. McTaggart, who denied the reality of the A-series, made the mistake of inferring from this the unreality of time; it is similarly tempting, and mistaken, to deny the importance of tensed thought in the way we must think about time. To say this is not to condone tensed thought as a kind of excusable, or psychologically inevitable, error. It is to say that when it is properly understood it is not an error at all. All thought and experience is undergone at particular moments in time, separate from points before and after, and this alone guarantees that certain indexical expressions which are couched in tensed terms are indispensable to proper thinking about time. It must, that is, be possible to refer to the point in time at which one is actually thinking, without knowing anything about the objective place of that moment in the B-series. If, therefore, advocates of tenseless time have succeeded in locating an error in the contrary opinion, it is the error of confusing indexical expression with expressions of some kind of 'objective' fact. There is no mistake in indexical thoughts, and there should be no suggestion that these thoughts are reducible in sense to non-indexical thoughts.

These familiar ideas, which in fact are stressed by noted advocates of tenseless time (such as Mellor), can help us to see another point, relevant to our moral concerns for the dead and the unborn. For I suggest that just as certain

---

[8] In talking of possible individuals I do not wish to imply any commitment to modal realism, but simply to refer to the results of genetic combinations, most of which have not been created.

indexical expressions are both indispensable and irreducible to proper thinking about time, so certain moral asymmetries may also be indispensable to
correct and intelligible moral thinking.

An immediate illustration of this is in retributive behaviour and the formation of moral attitudes towards others. Punishments, if justified at all, are
justified only on account of what people have done and not because of what
they will do. It may well be that a person's future misdemeanours have as
much objective existence and determinacy on the B-series as her past ones,
but even if we suppose that they do, this does not take away from the inap-
propriateness of punishing people for what they will do subsequently.[9] A central consideration here is that we typically foresee our future actions by directly
intending them (although perhaps we can also *predict* them by adopting a
third-person perspective on ourselves). But there seems to be no retrospect-
ive analogue of intention, no way in which I can intend to perform, or literally undo, some past deed of mine. That is why it can make sense only to
punish me for something I have done. To be punished for something I am
certainly going to do could only be understood as somehow justifying my
intention to do it. The oddity of all this is made clearer when we see that
what are often loosely known as 'forward-looking' theories of punishment
do not literally justify punishment on account of what a person is going to
do, but rather on account of the supposed influence of the punishment in pre-
venting future crime from being committed at all, and on the assumption that
some misdemeanour has already been committed. Moreover, what is true of
our attitudes to punishment has much more general application as well. For
our dealings with others are not primarily motivated by the thought that we
are dealing with abstractly conceived persons, but, on the contrary, with men
and women with particular characters. Our moral behaviour towards them is
guided by their being friends, lovers, people who share their past with us, or
to whom we owe certain things, given their past behaviour towards us. These
concrete facts about them are inevitably formed by past experience and events,
not by future ones. Even if future persons really exist somewhere further up
the metaphorical time-line, this is not enough to motivate the concrete sorts
of moral concern that we bestow on present and past persons. We are led by
knowledge of people's character and their concrete relationships with us, when
we decide on how to behave towards them.

---

[9] Or at least, in normal circumstances. One can imagine cases in which we know with near-
certainty that someone is about to commit some dreadful wrong, and cannot be prevented from
doing so, and we might feel inclined to punish him all the same. But here, it is plausible that
there must still be a present intention to commit the wrong, and that the punishment expresses
our abhorrence of the state of mind it embodies. It still seems inappropriate to punish people
for wrongs that they have as yet conceived no intention at all of committing.

One obvious objection might be aired here, stating that if we really take the tenseless conception seriously, we must concede that future persons are every bit as concrete in their characters and relationships as present and past ones are. This is, of course, true, but it is doubtfully relevant. The fact that causation runs (so we assume at least) from past to future[10] brings with it an asymmetry in our knowledge of past and future. This asymmetry is perhaps not logically necessary: one can imagine that the laws of nature were such that we possessed more detailed knowledge of the future than of the past. For example, we can imagine having very defective memories but excellent inductive powers of prediction. However, the asymmetry does in fact obtain, and nothing illustrates this better than our attitudes and behaviour towards the dead. We have already mentioned, by way of comparison with the unborn, some ways of tackling the problems posed by the non-existence of the dead, with respect to our obligations to them. But whatever the correct solution, it is clear that our concern for the dead is motivated by the fact that they died when their characters, wishes, and life narratives were determinate. Hence we feel obliged to respect particular desires they had, and mourn them precisely because we established particular relationships with them. None of this can be true of the not-yet-born. Even conceding the lack of ultimate, metaphysical difference between future and past, there is a crucial sense in which the past lives on in the present, whilst the future can be conceived as so many abstract possibilities. This is not affected by the supposition that there is some point even further on in the future, when people who (as we write) are as yet unborn will themselves be dead, and mourned as particular, formed individuals. For the point is that it is appropriate for those whose lives succeed them to regard them in this way, but not appropriate or even possible for those who precede them so to regard them.

I have tried throughout to consider these problems without committing myself on the issue of tensed versus tenseless time, asking instead what would be the implications for morality on each hypothesis. This non-committal stance may seem disingenuous. For what it is worth, it seems that the tenseless view is easier to argue for than the opposite view; it seems in any case that most of the things we naturally wish to say about time and tense can be said quite intelligibly without needing to suppose that tense is a genuine feature of reality. This itself is a strong consideration in favour of tenseless time, and is quite independent of the more complex claim, originating with McTaggart, that the tensed view is self-contradictory. My main aim, however, has been somewhat deflationary. I have tried to identify a number of moral considerations that

---

[10] I am simply assuming that causation runs forward in time and not backwards, and do not argue for this point here.

may seem to depend upon some stance about tense, and have suggested that the impact upon morality of these views on time is less than one might imagine.

We are still left, however, with a disturbing difficulty over how we are to understand the idea of obligations towards future people, in the light of the moral asymmetries mentioned, and in the light of the challenge, which seemed implicit in the tensed conception, that obligations are owed only (at most) to possible future individuals rather than exclusively to actual ones. The simple answer is that it is a mistake to think that, if there are moral obligations at all, these must be to specific, nameable persons. If the future is open and indeterminate, then I do not owe anything now to (say) my grandchild, conceived as a specific individual; but I can owe certain things to that person who will be my grandchild, whoever he or she will be. In a case like this, it is helpful to separate the role from the individual, understanding familial obligations as duties towards whoever fulfils these roles. Whether these roles will exist in future is clearly not certain—but since the point about the tensed, open-future theory is not to induce scepticism about the future, but to highlight difficulties about reference to future entities or events, any uncertainty about the future is not to the point. Similarly, we can be almost certain that several hundred years from now, there will be human beings on the planet— that is, that there will be future generations, even though the people who will collectively fulfil this role cannot be known, and perhaps are not yet determinate. It is thus important to care about the welfare of these future generations as much as we can, not only in material and ecological ways, but in the transmission of culture and knowledge, and all those things that will reinforce their sense of historical continuity with us. For whatever the problems of our conception of them, it is clear that they will have some determinate conception of us, as they build up a picture of their past. They will see themselves as produced and fashioned by us, and however vague our sense of what we owe to them, their sense of what they will have received from us is likely to be far clearer.

REFERENCES

Nagel, T. (1986), *The View from Nowhere* (New York: Oxford University Press).
Perrett, Roy (1987), *Death and Immortality* (Dordrecht: Martinus Nijhoff Publishers).
Silverstein, Harry (1980), 'The Evil of Death', *Journal of Philosophy*, 77: 7.
Yourgrau, Palle (1987), 'The Dead', *Journal of Philosophy*, 84/2: 84–101.

The Tensed vs. Tenseless Theory of Time: A Watershed for the Conception of Divine Eternity

WILLIAM LANE CRAIG

1. Introduction

Although theologians in the Judaeo-Christian tradition agree that God is eternal (without beginning or end of existence), they differ profoundly concerning the nature of divine eternity. The biblical data are not wholly determinative with respect to this question,[1] leaving theologians divided as to whether God's eternity is to be construed as timelessness or as infinite omnitemporality.

Some contemporary thinkers have claimed that the very idea of a timeless God is incoherent.[2] But it is not too difficult, I think, to conceive of God as existing changelessly *sans* the universe and, hence (on some relational view of time), timelessly. Since, in Judaeo-Christian understanding, creation is a freely willed act of God, there are possible worlds in which God exists alone, self-sufficient, without any created order. I have yet to find any good argument that in such a world God could not exist changelessly and timelessly.

The real difficulty arises, rather, when we consider worlds, like ours, in which both God and a temporal, created order exist. For given the reality of tense and God's causal relation to the world, it is very difficult to conceive how God could remain untouched by the world's temporality.

---

[1] See discussion and literature in Padgett (1992: 23–37); Helm (1988: 2–11). The biblical texts indisputably speak of God in temporal terms (e.g. Ps. 90: 2); but, given the anthropomorphic character of biblical language, one must be cautious in one's inferences. A key issue is whether the biblical authors had the reflective context required to entertain such a question. In certain places, God's creation of time itself does seem to be reflected upon (Jude 25; Gen. 1: 1–5), which would imply divine timelessness, at least *sans* creation, or his existence in some non-metric, metaphysical time.

[2] See Coburn (1963: 155); Jantzen (1983); Lucas (1989: 213); Gale (1991: 52). *Contra* see Pike (1970: 124); Mann (1983: 270); Helm (1988: 64–5); Yates (1990: 173–4); Leftow (1991*b*: 285–90). Frequently the question of God's existing in relation to realities *extra se* is conflated with the question of whether he can exist timelessly *a se*. For a relevant non-theistic analysis, see Walker (1978: 34–41).

Imagine God existing changelessly alone without creation, with a changeless and eternal determination to create a temporal world. Since God is omnipotent, his will is done, and a temporal world begins to exist. (We may lay aside
for now the question whether this beginning of a temporal creation would
require some additional act of intentionality or exercise of power other than
God's timeless determination.) Now in such a case, either God existed temporally prior to creation or he did not. If he did exist alone temporally prior to
creation, then God is not timeless, but temporal, and the question is settled.
Suppose, then, that God did not exist temporally prior to creation. In that
case he exists timelessly *sans* creation. But once time begins at the moment
of creation, God either becomes temporal in virtue of his real, causal relation to time and the world or else he exists as timelessly with creation as he
does *sans* creation. But this second alternative seems quite impossible. At
the first moment of time, God stands in a new relation in which he did not
stand before (since there was no *before*). We need not characterize this as a
change in God (perhaps change entails a 'before' and 'after' for an enduring subject), but this is a real, causal relation which is at that moment new
to God and which he does not have in the state of existing *sans* creation.
Even if the beginning of the temporal world is the result of a timeless voli-
tion of God, the fact that the world is not sempiternal but began to exist out
of nothing demonstrates that God acquires a new relation at the moment of
creation. At the moment of creation, God comes into the relation of *sustaining
the universe* or at the very least that of *coexisting with the universe*, relations
which he did not before have. Since he is free to refrain from creation, God
could have never stood in those relations; but in virtue of his decision to
create a temporal universe God comes into a relation with the temporal world
the moment the temporal world springs into being. As God successively
sustains each subsequent moment or event in being, he experiences the flow
of time and acquires a growing past, as each moment elapses. Hence, even
if God remains intrinsically changeless in creating the world, he none the
less undergoes an extrinsic, or relational, change, which, if he is not already
temporal prior to the moment of creation, draws him into time at that very
moment in virtue of his real relation to the temporal, changing universe. So
even if God is timeless *sans* creation, his free decision to create a temporal
world constitutes also a free decision on his part to enter into time and to
experience the reality of tense and temporal becoming.

Opponents of divine timelessness can therefore be understood as claiming
that

God is timeless   (1)

and

God is creatively active in the temporal world   (2) are broadly logically incompatible, on the basis of the necessary truth of

> If God is creatively active in the temporal world, God is really
> related to the temporal world (3)

and

> If God is really related to the temporal world, God is temporal. (4)

Since (2) is essential to Judaeo-Christian theism, (1) must be abandoned.

2. Aquinas's solution

(a) Denial of God's real relation to the world

The classic Thomistic response to the above argument against divine time-lessness is, remarkably, to deny (3). Aquinas tacitly agrees that if God were really related to the temporal world, then he would be temporal, as (4) affirms. As Liske points out, in Thomas's view relations between God and creatures, like God's being Lord, first begin to exist at that moment of time at which creatures come into being (Liske 1993: 224).[3] In the coming to be of creatures, then, certain relations accrue to God anew and thus, if these relations be real for God, he must be temporal in light of his undergoing extrinsic change, wholly apart from the question of whether God undergoes intrinsic change in creating the world. Contemporary philosophers have tended to overlook this fact, focusing the debate on Thomas's contention that God timelessly wills not merely his effects but also the times at which those effects appear in the temporal series (Thomas Aquinas, *Summa contra gentiles* 2. 35. 3–5).[4] But an examination of the context of Aquinas's remarks on this head reveals that his concern there is to explain how God can immutably will a temporal world without that world's always existing, that is, without its having an infinite past (ibid. 2. 32–8). Thomas's discussion of philosophical arguments for 'the eternity of the world' presupposes a construal of 'eternity' as only sempiternity, and thus he speaks of God's eternity even in terms that smack of temporality: 'Nothing, therefore, prevents our saying that God's action existed

---

[3] According to Liske, the reason Thomas resisted recognizing God's real relation to the world is that 'Obviously he feared that the mere temporal obtaining of a relation from God, if it is real, requires that God Himself must be temporal' (Liske 1993: 218).

[4] Focusing on the question of whether the act of creating involves intrinsic change in God's will or activity to the neglect of Thomas's position on God's real relation to the world are Davis (1983: 12–13) and Wierenga (1989: 198). Because he softens Aquinas's doctrine of no real, divine relations to the world to mean merely that God immutably causes the world, Yates is also forced to recur to this theme, yet without explaining how God's timelessness could be preserved in the face of his real relations with a changing, temporal world (Yates 1990: 142, 159–60). Incredibly, Harris thinks that 'Aquinas chose to ignore the whole problem' (Harris 1992: 74)! Similarly oblivious to Aquinas's solution is La Croix (1982: 391–9).

from all eternity, whereas its effect was not present from eternity, but existed at that time when, from all eternity, he ordained it' (ibid. 2. 35. 3. Cf. 2. 35. 5). Even if successful, Thomas's argument at best shows that God's efficacious will remains changeless as the world comes to be and as events successively occur and pass away. He says nothing in this context (nor was that his intention) to show how the origin and unfolding of a temporal world would not taint the eternal God with temporality in virtue of his real relation to the temporal sequence of events changelessly willed by him.

Aquinas's solution to the problem at hand is quite different: he denies that God has any real relation to the world. This prima-facie incredible position is rooted in Thomas's doctrine of divine simplicity, which is in turn based upon Aquinas's understanding of God as *ipsum esse subsistens*, the unrestricted act of being. In Aquinas's understanding, God does not have any nature or essence distinct from his act of existing (Thomas Aquinas, *Summa theologiae* 1a. 3. 4). For if a thing has an essence distinct from its being, it must have an existential cause which sustains it in existence. But God, as the Uncaused First Cause, cannot have a cause, and therefore his nature must be identical with his existence. Similarly, any thing having an essence distinct from its existing has by that fact the potentiality for existence. But since God, as the Unmoved First Mover, has no potentiality, his essence cannot be other than his existence. Now as the pure act of being, not defined by any essence, God is absolutely simple (ibid. 1a. 3. 1–7). From God's simplicity and the utter absence in him of any potentiality, God's immutability follows (ibid. 1a. 9. 1), and it is on the basis of God's immutability that Thomas infers God's timeless eternity (ibid. 1a. 10. 2. Cf. Aquinas, *Summa contra gentiles* 1. 15).

It is at this point that our objection arises: even if God immutably wills the creation of the temporal world, would not the origin of that world, in virtue of God's relation to it, bring God into time? Thomas has already implicitly invalidated such a question in his doctrine of divine simplicity. For God's being simple entails, in particular, that God transcends the Aristotelian metaphysical distinction between a substance and its accidents. For Aquinas accidents are properties which a thing, or substance, may possess either contingently or necessarily, but which do not enter into the definition of what the thing is, or its essence. Thomas bases his denial of accidents in God squarely on his conception of God as being itself or pure actuality (Aquinas, *Summa theologiae* 1a. 3. 6). The importance of the absence of accidents in God becomes evident when we recall that one of the nine categories of accident listed by Aristotle was relation ($\pi\rho\grave{o}\varsigma$ $\tau\iota$). According to Thomas's peculiarly Aristotelian metaphysic, relations are actually monadic predicates or properties inhering in one or both of the *relata*. Though a relation might be grammatically or logically predicated of both *relata*, the ontological accident of relation might not inhere in both terms of that relation. Aquinas distinguished three possibilities in this regard: (1) the relation may exist merely in thought, not in the things themselves, as is the case with the relation of self-identity; (2) the relation may exist in both things, as in relations of quantity; (3) the relation may exist in one *relatum* only, being purely ideal for the other, as in the case of a knower and the object known (ibid. 1a. 13. 7). In this third case, the relational predicate signifies something real (*res naturae*) in the one *relatum*, but only something conceptual (*res rationis*) for the other. A knower K has the real property of *knowing object O*, but O itself does not possess any real property of *being known by K*, as is evident from the fact that O would be intrinsically the same if K were non-existent, whereas K would be intrinsically different were O not to exist.

Now since God is simple and lacks all accidents, he cannot possess any relations to creatures. Therefore, according to Aquinas, while the temporal world does have the real relation of *being sustained by God*, God does not have a real relation of *sustaining the temporal world*. This latter relation, while predicated of God, in fact signifies only a conceptual relation. Startling as it may sound, God does not have any relations of Creator to creature, cause to effect, Saviour to saved, and so forth. Aquinas writes,

> Whenever two things are related to each other in such a way that one depends upon the other but the other does not depend upon it, there is a real relation in the dependent member, but in the independent member the relation is merely one of reason—simply because one thing cannot be understood as being related to it. The notion of such a relation becomes clear if we consider knowledge, which depends on what is known, although the latter does not depend on it.
>
> Consequently, since all creatures depend on God, but he does not depend on them, there are real relations in creatures, referring them to God. The opposite relations in God to creatures, however, are merely conceptual relations; but, because names are signs of concepts, certain names we use for God imply a relation to creatures, even though, as we have said, this relation is merely conceptual. (Thomas Aquinas, *De veritate* 4. 5. Cf. *Summa theologiae* 1a. 13. 7; *Summa contra gentiles* 2. 11–14; *De potentia Dei* 3. 3)

The fact that relations between God and creatures inhere only in the latter enables Aquinas to avert the objection to divine timelessness based on God's relation to the temporal world. He explains,

> whatever receives something anew must be changed, either essentially or accidentally. Now certain relations are predicated of God anew; for example, that he is Lord or governor of this thing which begins to exist anew. Hence, if a relation were predicated of God as really existing in him, it would follow that something accrues to God anew, and thus that he is changed either essentially or accidentally; the contrary of this having been proved . . . (Aquinas, *Summa contra gentiles* 2. 12. 5)

Since God is immutable, the new relations predicated of him at the moment of creation are just in our minds; in reality the temporal world itself is created with a relation inhering in it of *dependence on God*. Hence, God's timelessness is not jeopardized by his creation of a temporal world.

This unusual doctrine of creation becomes even stranger when we reflect on the fact that creating a temporal universe is an act of God and that action, like relation, is one of the nine Aristotelian categories. It would seem to follow that God has no real actions and therefore cannot properly be said to have created the world (though the world could have under the category of passivity or passion the accident of being created by God). Aquinas escapes this conclusion, however, by identifying God's action with his power and, hence, with his essence (ibid. 2. 9. 5). God's act of being is his power and his act of creating. Thus, in creating the world God does not perform some act extrinsic to his nature; rather the creature (which undergoes no change but simply begins to exist) begins to be with a relation to God of *being created by God*:

Taken actively, [creation] denotes the act of God, which is his essence, together with a relation to the creature: and this is not a real but only a logical relation. But taken passively, since . . . it is not properly speaking a change, it must be said to belong, not to the genus of passion, but to that of relation. . . . Creation taken actively denotes the divine action to which the mind attaches a certain relation . . .: but taken passively, . . . it is a real relation signified after the manner of a change on account of the newness or beginning that it implies. (Aquinas, *De potentia Dei* 3. 3)[5]

According to this doctrine, then, God in freely creating the universe does not really do anything different from what he would have done, had he refrained from creating; the only difference is to be found in the universe itself: instead of God existing alone *sans* the universe we have a universe springing into being at the first moment of time possessing the property *being sustained by God*, even though God, for his part, bears no real reciprocal relation to the universe made by him.

(*b*) *Assessment of Thomas's solution*

By way of assessment, I think it hardly needs to be said that Thomas's solution, despite its daring and ingenuity, is extraordinarily implausible. Wholly apart from the very problematic notions of God's essence being identical with his act of being and of God's simplicity, we have this very difficult tenet that while creatures are really related to God, God is not really related to

---

[5] Cf. his comment, 'Consequently creation is really nothing but a relation of the creature to the Creator together with a beginning of existence.'

creatures. How are we to make sense of this idea? For Aquinas a real relation is one that obtains objectively in the real world; a mental or conceptual
relation is one posited by the mind, but having no counterpart in reality.
Analogously, the distinction between God's will and his existence is not real,
but conceptual; or again, we can imagine God prior to the moment of creation, but really there was no such prior time (Aquinas, *Summa theologiae*
1a. 19. 2; *De potentia Dei* 3. 1. 2). Now a real relation is for Aquinas a property inhering in a substance. This understanding may seem strange to us, since
we normally conceive of relations as being polyadically, not monadically,
predicated. As we conceive relations, it would seem impossible for a real
relation to obtain between two things without that relation being real for both
of them. None the less, Aquinas does seem to be onto something important
in distinguishing real from conceptual relations. In certain cases, the foundation of a relation between two things is constituted by the intrinsic properties
of only one of the *relata*. For example, if I resent my boss, then I stand in
a *resentful of* relation to him, and he stands in a *resented by* relation to me.
But the foundation of these reciprocal relations lies in my intrinsic properties,
not in those of my boss. This is not to say that my boss has done nothing to
cause or merit my resentment; it is simply to say that the relation itself obtains
wholly because of intrinsic properties I possess, regardless of the source of
those feelings. So in a sense, a relation can be said to be asymmetrically real
if it is founded on intrinsic properties of only one of its *relata*. Perhaps in
such a case we could say that the *relatum* on whose intrinsic properties the
relation is founded has a real, intrinsic, relational property, for example *resent-
ing Jones*, but that the other *relatum* possesses no real, intrinsic, relational
property like *resented by Smith*. Hence, it makes sense to say that among
certain *relata*, not all really possess intrinsic, relational properties, though all
stand in real relations to one another. Such monadically predicated properties would come close to what Aquinas understood by relations as accidents
inhering in a substance.

The question then is whether our predicating of God at the moment of creation the relational property of *sustaining the world* is merely conceptual or
ascribes a real property to him. 'Sustaining' clearly describes a relation which
is founded on something's intrinsic properties concerning its causal activity,
and therefore *sustaining the world* ought to be regarded as a real property
acquired by God at the moment of creation. I must confess that I find
Aquinas's position, that this property is not really possessed by God, but that
the relevant real, relational property is *being sustained by God*, which is
possessed by the world, to be quite incredible. If at the moment of creation
the world begins to exist with the relational property *being sustained by
God*, then how could God fail to acquire at that very moment the relational

property *sustaining the world*? Aquinas's own examples seem to betray him here. In the cases of knowledge and perception and their objects, the real relation is said to inhere in the person knowing and perceiving, not in the objects known and perceived. But surely God, as the Creator and Sustainer of the cosmos, is more analogous to the person knowing and perceiving than to the objects of his knowledge and perception. It seems fantastic to think that the relation between a cause and its effect is analogous to relations like *known by* or *envied by*. If the relation of some cause to its effect is unreal, then the cause has in particular no causal relation to its effect; that is to say, the cause is not a cause, which is self-contradictory. All we can say in such a case is that the effect is really related to another object or event as the effect of said object or event. In truth there is no real cause in such a case, only a real effect. But it seems unintelligible, if not contradictory, to say that one can have real effects without real causes. Yet this is precisely what Aquinas affirms with respect to God and the world. Words like 'First Cause' and 'Creator' are only extrinsic denominations applied to God, that is, predicates which do not correspond to any real property but which are appropriate in virtue of real properties in creatures. Even if we adopt the Thomist view that causation takes place entirely in the effect, not in the cause, that only underscores the reality of God's causal relation to the world, since the world is admitted to be really related to God as effect to cause, to be really caused by God, which is all that there is to causality; nothing more needs to be added *ex parte Dei* for him to be the cause of the world. Yet Thomism denies that God is literally the cause of the world, though the world is the effect of God—which seems contradictory or meaningless.[6]

A number of contemporary thinkers have sought to defend the Thomistic solution to the objection under consideration by claiming that a necessary condition of a real relation is some intrinsic *change* in the subject having that relation. Since the world's beginning to exist is said to be immutably and timelessly willed by God, its coming to be involves no intrinsic change in God and hence no real relation on God's part to the world. Thus, Peter Geach asserts that the denial that God is really related to the world is

---

[6] The fact that we are dealing with a causal relationship between God and the world makes the present objection to divine timelessness much more powerful than similar arguments by Wolterstorff or Smith for the temporal existence of God or abstract objects on the basis of changing reference to them by temporal agents (Wolterstorff 1975: 186–7; Smith 1993: 204–14). For clearly, relational properties like *referred to by Jones* are much more akin to relations like *envied by* than are relational properties like *sustaining the world* and are therefore more plausibly regarded as merely conceptual, not real. In defence of God's timeless creation of a temporal world, McCann (1993: 237–9) and Yates (1990: ch. 5) at best show that God's act of creating need not be an intrinsic change in him, but only an extrinsic change. They fail to show that an extrinsic change would not suffice to temporalize God.

traditionally bound up with the denial that God undergoes change. Contrasting real change to pseudo-change, or what he facetiously calls 'Cambridge change', Geach (1969: 71–2) takes God's becoming Creator to be merely a 'Cambridge change' for him (Geach 1972: 322–3; see also id. 1969: ch. 6). Geach has no criterion for discerning real change, and the examples of Cambridge change which he offers are instances of relational changes in objects undergoing no change of intrinsic, non-relational properties. Presumably, then, God at most changes extrinsically in creating a temporal world and so is not really related to the world.

Such reasoning is predicated on an incorrect understanding of real relations. Intrinsic change in a thing's properties is neither sufficient nor necessary for that thing's relation to something else being real. Thomas's paradigm example of an asymmetric real relation, a knower's relation to the object of knowledge, not only implies no intrinsic change, but no extrinsic change either; indeed, it could be a timeless and immutable relation. On the other hand, if the object of knowledge did undergo intrinsic change, that would do nothing to make its relation to the knower real. Similarly, creation itself is not, in Thomas's lights, any change in the thing created, but a sheer beginning to exist with a real relation of dependence on God. If intrinsic change were a necessary condition of real relations, then God and the event of creation do not stand in any real relation at all, whether from the side of the creature or of God, which is absurd.

The immutability of God's will, knowledge, and love in relation to creatures is thus wholly beside the point with respect to the question of God's real relation to the world. The issue is not intrinsic *change*, but intrinsic (counterfactual) *difference*: if a world of other creatures were actual, would God's will, knowledge, and love relationships be different? If we affirm this, then God has different intrinsic properties from world to world and so real relations with the creatures willed, known, and loved by him. To be sure, God's being different in will, knowledge, and love across various possible worlds is ultimately due to his own free decree as to what sort of creatures to create. But the dependence of which creatures are actual upon God only shows that God's relation to creatures is freely chosen by him, not that that relation is unreal (see Hill 1974: 157).

Wholly apart from the question of intrinsic change on God's part, then, the admission that God is intrinsically different in different possible worlds, in that what he knows, wills, and loves is diverse across worlds, demonstrates that his relation with creatures is not merely conceptual, all the diversity residing in the creatures alone, but real because it is founded in intrinsic properties of God himself. But if God has real relations with the temporal universe, no reason remains for denying God's temporality, even if his becoming Creator is a 'Cambridge' change. For even extrinsic change can be sufficient for real relations. In Geach's example of Socrates' becoming shorter than Theaetetus due to the latter's growth, only Theaetetus undergoes intrinsic change, but Socrates' being shorter than Theaetetus as a result of that change is still a real relation. With respect to creation, we have conceded for argument's sake that God's creating the world is not the consequence of an intrinsic change on his part. Accordingly, his becoming Creator could be construed as a 'Cambridge' change, resulting from the universe's springing into being. But it does not follow that the relation which accrues to God as a result is therefore unreal, since intrinsic change is not a necessary condition of a real relation. Even if God is conceived to be timeless *sans* creation, so that he cannot properly be said to *change* (even extrinsically) in virtue of the new relation he acquires at the first moment of time, still the newness of that relation suffices to bring him into time.

In fact, Aquinas himself makes no appeal to so-called 'Cambridge change', for he has a quite different way of eluding the dilemma of God's knowledge, will, and love's either being identical with his essence, thereby removing divine freedom and contingency, or else being accidental to him, thereby destroying divine simplicity and his unrelatedness to the world. Bizarre as it may sound, it is the implication of Aquinas's position that God is perfectly similar across possible worlds, even the same in worlds in which he refrains from creation as in worlds in which he creates. For in none of these worlds does God have any relation to anything *extra se*. In all these worlds God never acts differently, he never cognizes differently, he never wills differently; he is just the simple, unrelated act of being.

Of course, in these various worlds different creatures have the accidents of being sustained, known, and loved by God. The entire difference between worlds is to be found there, on the side of creatures. But that brings us back to Aquinas's doctrine of creation, which I previously characterized as unusually strange. In every world God exists in every respect the same. Even in worlds in which he does not create, his act of being, by which creation is produced, is no different in these otherwise empty worlds from what it is in worlds chock-full of contingent beings of every order. The only difference is that in worlds in which God creates there is, from God's perspective, that *relatio rationis* to finite things.

Unfortunately, Thomas's doctrine of creation is just not credible. Thomas's doctrine makes it unintelligible why the universe exists rather than nothing. The reason obviously cannot lie in God, either in his nature or his activity (which are only conceptually distinct anyway), for these are perfectly similar in every possible world. Nor can the reason lie in the creatures themselves, in that they have a real relation to God of *being freely willed by God*. For

their existing with that relation cannot be explanatorily prior to their existing with that relation. What is wanted is something posterior to God in the
order of explanation but prior to the existence of creatures really related to
God. But in Thomas's system there is an explanatory lacuna in that middle
position.

Few contemporary interpreters of St Thomas have faced this issue squarely.
John Wright does, and he finds himself forced to conclude that

we can't say that 'Creator' is wholly and simply a matter of extrinsic denomination
founded on the reality of an extrinsic denomination, that is, of creatures. It will not
do because the reality of creatures and of their dependence presupposes, not merely
logically but ontologically, the activity of God as determined to produce creatures
and to produce these rather than some other possible creatures. We may call this determination what we like, but we cannot reduce it merely to a posterior construction of
the human mind. To do so would be to make the actual existence of the world either
absurd or independent of God (since then there is objectively nothing in the divine
activity, no reason at all why creatures exist rather than not exist, or these creatures
rather than some other possible ones) or else to make it the inevitable consequence
of necessary divine activity. (Wright 1977: 457)

Making the existence of the universe the inevitable consequence of divine
activity results from saying that the reason the universe exists is due to the
simple essence of God; making the existence of the universe independent of
God results from saying that there is no reason why creatures exist rather
than nothing; and making the existence of the universe absurd results from
Thomism, saying that God has no real relation to the world, but the world
has a real relation to God.

I conclude, then, that the escape from the present objection advocated by
Thomists, namely, denying the truth of (3), ultimately leads to absurdity and
so must be rejected.

3. Stump and Kretzmann's solution

(a) ET-simultaneity

If, then, we agree that God is really related to the world in virtue of his creative activity in the temporal world, then we must deny the necessary truth
of (4) in order to undercut the argument at issue for divine temporality. In
1981 Eleonore Stump and Norman Kretzmann sparked a renewal of interest
in the doctrine of divine timelessness by proposing a model of God's relationship to time which allegedly demonstrated the possibility of God's being
atemporal and yet really related to the world.

The heart of the Stump–Kretzmann proposal lies in their conception of a new species of simultaneity, which they call 'eternal-temporal simultaneity' (or 'ET-simultaneity') (Stump and Kretzmann 1981: 434–40). They take the generic concept of simultaneity to be *existence or occurrence at once* (that is, *together*). 'Temporal simultaneity' refers to a species of this generic concept and means *existence or occurrence at one and the same time*. Temporal simultaneity and simultaneity are not the same, since between two eternal entities or events there obtains another species of the generic concept of simultaneity called 'eternal simultaneity', which is *existence or occurrence at one and the same eternal present*. Thus, the two species of simultaneity are distinguished by the specific content given to the general notion *at once*, or *together*. Simultaneity in general involves coexistence or co-occurrence, but does not specify whether this coexistence or co-occurrence is at one and the same time or at one and the same eternal present.

The problem seen by Stump and Kretzmann in relating an eternal entity to a temporal entity is that there is no single mode of existence shared by the two entities. Hence, one cannot draft a formulation of ET-simultaneity on the usual pattern 'existence or occurrence at one and the same ——'. 'What is temporal and what is eternal can co-exist . . . but not within the same mode of existence and there is no single mode of existence that can be referred to in filling in the blank in such a definition of ET-simultaneity' (ibid. 437–8).

At this point Stump and Kretzmann turn to the Special Theory of Relativity (STR) in order to provide an analogy to the type of simultaneity relation they will propose. In STR simultaneity is redefined as *existence or occurrence at the same time within the reference frame of a given observer*. This conception of simultaneity relative to a reference frame suggests a way to define ET-simultaneity. Stump and Kretzmann propose that we construe a mode of existence as analogous to a reference frame and construct a definition in terms of two reference frames and two observers.

(b) First attempt at a definition of 'ET-simultaneity'

Accordingly, Stump and Kretzmann formulate the following definition of ET-simultaneity:

For every x and for every y, x and y are ET-simultaneous iff
  (i) either x is eternal and y is temporal, or vice versa; and
  (ii) for some observer, A, in the unique eternal reference frame, x and y are both present—i.e., either x is eternally present and y is observed as temporally present, or vice versa; and
  (iii) for some observer, B, in one of the infinitely many temporal reference frames, x and y are both present—i.e., either x is observed as eternally present and y is temporally present, or vice versa. (Ibid. 441)

The first thing to notice about this definition is that it is not really analogous to simultaneity in STR at all. In STR two simultaneous events both occur at one and the same time relative to a given reference frame. The analogy to this would be that two events (or entities), one eternal and one temporal, are simultaneous iff relative to the eternal 'reference frame' both occur at one and the same eternal point and relative to the temporal 'reference frame' both occur at one and the same time. ET-simultaneity as defined by Stump and Kretzmann only remotely resembles relativistic simultaneity in that two 'reference frames' are employed. But beyond that it is quite non-analogous to simultaneity as defined in STR.

A further clarification of clause (iii) is in order. When the definition refers to 'the infinitely many temporal reference frames', it might naturally be thought that physical reference frames in STR are being referred to. But this would contradict Stump and Kretzmann's repeated statements that their account presupposes no more than a Newtonian, absolute time and that STR serves as merely an introductory analogue to their definition of ET-simultaneity. Their definition utilizes 'reference frame' metaphorically to refer on the one hand to God's atemporal mode of existence and on the other hand to our temporal mode of existence. Thus, the 'infinitely many temporal reference frames' refer to the different moments of time. A reference frame is thus not strictly analogous to a mode of existence, but rather to locations on a geometrical representation of a mode of existence.

It is noteworthy that in the proposed definition simultaneity is not defined in terms of a shared location, but in terms of a shared property. Relative to a location either in time or in eternity, both x and y are said to be present. This is not a shared location (contrast: 'in the present'), since x and y are not both located in the 'eternal present' nor in any temporal present. Such a procedure seems peculiar, since two entities' sharing a property relative to some location hardly suffices for simultaneity. Relative to the eternal reference frame, for example, God and Jones are both intelligent, but they are not therefore in any way simultaneous. But when it comes to the property of presentness, I think, we can make sense of such a procedure. For example, we could define temporal simultaneity by stating that x and y are simultaneous iff, relative to some time t, x and y are both present. The problem with the Stump–Kretzmann definition is that the word 'present' in the definition refers to entirely different properties, namely, temporal presentness and eternal presentness, so that there is no shared property involved. We cannot circumvent this problem by giving tenseless, token-reflexive truth-conditions relative to eternity or to moments of time for statements like 'y is present', since in eternity as well as at most moments of time there are no such tokens. Rather we must find some common property shared by God and temporal entities relative to either's 'reference frame' which intuitively suffices to found a simultaneity relation. I think that the essence of the Stump–Kretzmann definition would be preserved if we state that relative to either frame 'x and y are both real', one eternally real and the other observed as temporally real relative to the eternal 'reference frame' or one temporally real and the other observed as eternally real relative to a moment of time.

On the basis of their definition of ET-simultaneity, Stump and Kretzmann believe themselves to have solved the problem of a timeless deity's being active in the world. In virtue of their ET-simultaneity God and events in time are coexistent:

if anything exists eternally, its existence, although infinitely extended, is fully realized, all present at once. Thus the entire life of any eternal entity is co-existent with any temporal entity at any time at which that temporal entity exists. From a temporal stand-point, the present is ET-simultaneous with the whole infinite extent of an eternal entity's life. From the standpoint of eternity, every time is present, co-occurrent with the whole of infinite atemporal duration. (Ibid.)

As an illustration of this coexistence, Stump and Kretzmann invite us to im-agine two infinite, parallel, horizontal strips, the upper one (representing eter-nity) being a strip of light (light representing the present) and the lower one (representing time) being dark except for a dot of light moving steadily along it. 'For any instant of time as that instant is present, the whole of eternity is present at once; the infinitely enduring, indivisible eternal present is simul-taneous with each temporal instant as it is the present instant' (Stump and Kretzmann 1987: 219).

Whatever one may think of their doctrine, this illustration is plainly con-fused, mixing as it does spatial and temporal imagery. From the standpoint of eternity, the eternal 'present' is wholly simultaneous with each instant *as it becomes present* (illuminated). Thus, eternity is simultaneous with each moment of time in succession; otherwise, from the standpoint of eternity the lower strip would have to be wholly illuminated, like the upper strip. Given that the eternal 'present' is successively simultaneous with one temporal pres-ent at a time, eternity is not atemporal at all, but has been temporalized in virtue of its real relation to time. In other words, Stump and Kretzmann's illustration portrays vividly precisely the objection currently under consid-eration, which their account is supposed to resolve.

A more apt illustration of their view would have drawn upon the relativity of simultaneity: from the standpoint of any temporal present, the lower strip is a single dot of light at that point and the upper strip is observed as wholly illuminated, and from the standpoint of eternity, eternity is a single, indivis-ible point of light and the entire temporal strip is observed as illuminated.

The upshot of the doctrine of ET-simultaneity is that all temporal events are observed as present (or, as I have suggested, real) by God. Stump and Kretzmann make it quite clear that by the expression 'observed as present' they do not mean merely 'epistemically present' to God, that is, 'gathered into one specious present' by God. Rather all temporal events are onto-logically real for God. Stump and Kretzmann emphasize that this doctrine does not imply that temporal events exist in eternity or that tense and temporal becoming are illusory; rather it implies a sort of 'metaphysical relativism' (Stump and Kretzmann 1981: 442–3).[7] Reality is composed of two incom-mensurable modes of real existence, time and eternity, which cannot be brought together into a single frame of reference. God veridically experiences every temporal event as having presentness in relation to eternity.

Perhaps we can get a better understanding of Stump and Kretzmann's view by drawing once more upon relativity theory: just as at a single spacetime location different observers will observe different simultaneity classes of events depending upon each observer's velocity at that point and none of these can claim to be preferred as the uniquely correct simultaneity class, so at the point of eternity there is no unique, preferred class of temporally simultaneous events which are observed as real and so temporally present by God, but God observes different classes of temporally simultaneous events to be real in accordance with the ET-simultaneity relation. The disanalogy in the eternity/temporality case is, of course, that instead of a plurality of observers we have only one observer, God, and there is no ground, such as a physical observer's vel-ocity and the constancy of light's speed, of this metaphysical relativity.

The radical metaphysical relativity postulated by ET-simultaneity implies that all events in time are present or real to God in eternity and therefore open to his timeless causal influence. 'Even though His *actions* cannot be located in time, he can bring about *effects* in time' (1981: 448 (my empha-sis)). Every action of God is ET-simultaneous with any temporal effect ascribed to it, and 'ET-simultaneity is a sufficient condition for the possibility of a causal connection in the case of God's bringing about the existence of a tem-poral entity' (ibid. 451). Thus, God's causal relation to the world is grounded in his being ET-simultaneous with every event in time.

Now in order to be ET-simultaneous with some temporal event *y*, *x* must be atemporally real and *y* be observed to be temporally real. But, as many critics have pointed out, the language of observation in the definition of ET-simultaneity is wholly obscure.[8] In STR physical operations involving clock

---

[7] They should have said 'relativity', I think.

[8] Davis (1983: 20); Lewis (1984: 74–6); Helm (1988: 32–3); Hasker (1989: 164–6); Yates (1990: 128–30); Leftow (1991*b*: 170–2). Unfortunately, many of these critics, misunderstanding the role of hypothetical observers in relativity theory, think that Stump and Kretzmann require

synchronization via light signals are stipulated in order to give physical meaning to simultaneity at a distance. When y is 'observed' to be simultaneous with x according to STR, the word 'observed' might more perspicuously be replaced with 'calculated'; the determination of y's simultaneity with x is based not on physical observation of y, but on the solution to a mathematical equation. But in the definition of ET-simultaneity no hint is given as to what is meant, for example, by x's being observed as eternally present relative to a time t. In the absence of any procedure for determining ET-simultaneity, the language of observation becomes vacuous. All that is meant is that relative to the 'reference frame' of eternity x is eternally present (or real) and y is temporally present (or real) and that relative to some temporal 'reference frame' x is eternally present (or real) any y is temporally present (or real). It hardly needs to be said that such a definition clarifies nothing. It only restates the fact that x is atemporal and y is temporal without explaining in what sense they are simultaneous.

Nor will an appeal to metaphysical relativity help to make sense of the definition. In the absence of operational definitions which serve to redefine distant simultaneity in STR, there really are no simultaneous events at all at a distance. Absent these operational definitions, there just is no fact of the matter concerning the simultaneity of spatially separated events. Only events at the same spacetime point are simultaneous. Thus, if reality is bifurcated into two irreducible modes of existence, and there is no way of observing or determining what is ET-simultaneous with x or y, there just is no ET-simultaneity between them. Only events co-occurring at the same time or at the same eternal point are simultaneous, that is to say, only temporal and eternal simultaneity obtain. It follows that an atemporal God could not be causally related to the world.

Moreover, even if we grant metaphysical relativity on the analogy of temporal relativity, it will not yield ET-simultaneity. To say that y is observed by God as temporally present or real just means that relative to God y is temporally present or real. But if $w < y < z$, then in so far as y is temporally present to God w is past and z is future. Of course, since, as McTaggart insisted, all events are eventually present, it follows that due to metaphysical relativity w is also temporally present relative to God in another metaphysical frame of reference, as is z in yet another metaphysical frame of reference. But in so far as we consider reality from the perspective in which God is eternal

that a temporal person somehow actually observe God as eternally present, which is impossible. But Stump and Kretzmann are clear that for them an observer is anything with respect to which a reference frame is determined (Stump and Kretzmann 1981: 438; 1992: 474). Actually Stump and Kretzmann concede too much, for all that is required in relativity theory is hypothetical observers.

and y is temporally present relative to God, there are events earlier and later than y which are past and future respectively relative to God. But if y is temporally present relative to God, while w is past and z is future relative to God, then God and y are temporally simultaneous, not ET-simultaneous. Thus, the sort of simultaneity suggested by metaphysical relativity is not ET-simultaneity, but an extension of physical relativity of simultaneity into metaphysics, that is to say, God is temporally simultaneous with every temporal present. From our perspective, these temporal presents are evanescent, but from God's multitude of perspectives each present is simultaneous with him in some perspective on reality. Such a view obviously fails to preserve God's timelessness.

No doubt Stump and Kretzmann would cry foul at such a critique, in that I have pushed the analogy to relativity theory far beyond their intent. I concede the point; but then I simply cannot make sense out of the language of observation found in their definition nor of the metaphysical relativity appealed to in its explication. Stump and Kretzmann are not really metaphysical relativists, but hold that God has a unique perspective on the world according to which all events are in some unexplained way equally real to him.

Given, then, that Stump and Kretzmann's definition of ET-simultaneity is merely a non-explanatory restatement of the doctrine that while God is timeless all things in time are temporally present or real to him, it cannot do the explanatory work necessary to show why

> If God is really related to the temporal world, God is temporal   (4)

is not necessarily true.

(c) Second attempt at a definition of 'ET-simultaneity'

Fortunately, Stump and Kretzmann have for theological reasons tried to free their definition from observation language, and perhaps this will give some explanatory content to the definition of ET-simultaneity. They now propose:

For every x and every y, x and y are ET-simultaneous if and only if
(i) either x is eternal and y is temporal, or vice versa (for convenience, let x be eternal and y temporal); and
(ii) with respect to some A in the unique eternal reference frame, x and y are both present—i.e., (a) x is in the eternal present with respect to A, (b) y is in the temporal present, and (c) both x and y are situated with respect to A in such a way that A can enter into direct and immediate causal relations with each of them and (if capable of awareness) can be directly aware of each of them; and
(iii) with respect to some B in one of the infinitely many temporal reference frames, x and y are both present—i.e., (a) x is in the eternal present, (b) y is at the

same time as B, and (c) both x and y are situated with respect to B in such a way that B can enter into direct and immediate causal relations with each of them and (if capable of awareness) can be directly aware of each of them. (Stump and Kretzmann 1992: 477–8)

This new definition is quite strange. Although the observation-words are absent, the two observers A and B remain. Since the role in the original definition of A and B was simply to specify 'reference frames', one might surmise that we could simply eliminate them altogether and just label the eternal 'frame' A and the temporal 'frame' B, so that relative to the eternal 'reference frame' A, x is in the eternal 'present' and y is in the temporal present—but then we find that clause (ii)(c) requires that A itself enter into causal relations, which a mere reference frame cannot do. A and B have in this redefinition become real, causally efficacious beings. Moreover, since God alone is eternal, $A = x =$ God. It follows that according to (ii)(c) God is so situated in respect to himself that he can enter into causal relations with himself. God must be in some sense self-caused. Furthermore, unless Stump and Kretzmann are willing to countenance creatures in some way causally influencing God, (iii)(c) must imply that the creature B can be merely a direct and immediate effect of God.

   Another curious feature and serious drawback of this definition is that the only temporal events which are ET-simultaneous with God turn out to be present events. For relative to eternity, (ii)(b) stipulates that y must be in the temporal present, not just at some moment of time. Thus, from God's perspective only the temporal moment which now exists is ET-simultaneous with him. Since (iii)(b) is not parallel to (ii)(b) in that, unlike the latter, it refers not to the temporal present, but to any moment in time, it might be thought that relative to any moment of time a temporal event at that time is ET-simultaneous with God. But if the verbs 'can enter' and 'can be' in (iii)(c) are tensed, then again only present events are ET-simultaneous with God, since B can (presently) enter into causal relations only if B exists now.

   God's being ET-simultaneous with only present events might appear at first blush to be acceptable to the theorist of tensed time, who holds to metaphysical presentism and regards past and future things/events as non-existent. One cannot, after all, be simultaneous with non-existent entities, so perhaps God's being ET-simultaneous with present events alone is not so bad. Stump and Kretzmann would probably not welcome the consequence that their theory is incompatible with the tenseless theory of time, however. And a moment's reflection reveals that God's being ET-simultaneous with only present events leads to incoherence. For relative to God, the only events he is ET-simultaneous with are present events. But since which events are present is constantly changing, God acquires continually new relations of ET-simultaneity. Hence, he is changing and therefore temporal. Thus, if God is simultaneous only with present events, his relation to them is ordinary temporal simultaneity, not ET-simultaneity.

Hence, Stump and Kretzmann's new definition needs to be read tenselessly throughout, including clause (iii)(*c*), and (ii)(*b*) must be amended to state '*y* is at some point in time'. So amended, the definition successfully stipulates that God is ET-simultaneous with every temporal event, whether it be past, present, or future.

But now we come to what appears to be an irremediable problem with the new definition. Notice that although ET-simultaneity is still defined in terms of a shared property of presentness, presentness is now explicated in terms of different locations and a shared property of being situated in a certain way. Thus, unlike the first definition, this redefinition does provide a univocal sense in which God and creatures are said to be present, namely, that although they are differently located, in time and eternity, they can enter into direct and immediate causal relations and be directly aware of each other.

The problem with this new definition of ET-simultaneity, however, is that it makes Stump and Kretzmann's account of divine eternity viciously circular, as Leftow has charged. For ET-simultaneity was originally invoked in order to explain how a timeless deity could be causally active in time, but now ET-simultaneity is defined in terms of a timeless being's ability to be causally active in time. As Leftow states,

any definition of ET-simultaneity which invokes any form of ET-causality (or . . . other causally implicated ET-knowledge) is implicitly circular. For to fully explain how ET-causation can occur, we must bring in the concept of ET-simultaneity. If we do, we cannot then define ET-simultaneity by invoking ET-causation, for then the concept to be defined in effect recurs in the definition. (Leftow 1991*b*: 173)

The new definition proffered by Stump and Kretzmann (once (ii)(*b*) is amended) simply takes it for granted that God and temporal events are so situated (whether with respect to God or to events) that God can be directly and immediately causally related to them while remaining atemporal. But the objection which drives our present concern is precisely that a timeless God and temporal events cannot be so situated. How can God be really related to a present event without himself being present? The answer is supposed to be that they are ET-simultaneous. But now ET-simultaneity is explicated in terms of God's ability to be causally related to temporal events while remaining atemporal. Such an account will remain viciously circular unless and until Stump and Kretzmann provide an independent explanation of how the timeless God can be directly and immediately causally related to events in time.

Thus it seems to me that Stump and Kretzmann have failed to provide any coherent model which explains how God can be both atemporal and yet causally active in the world. Their first definition of so-called ET-simultaneity was explanatorily vacuous, a mere restatement of the fact that God is atemporal and temporal events are present. Their redefinition explicated ET-simultaneity in terms of causal and epistemic relations, which they had previously tried to explain in terms of ET-simultaneity, thus closing a vicious circle. In short, they have said nothing which would undercut the argument that in virtue of his real relation to the temporal world God is temporal.

4. Leftow's solution

(a) The existence of all things in eternity

The central difficulty and in the end fatal flaw of Stump and Kretzmann's attempt to enunciate an informative doctrine of ET-simultaneity is their inability to bring together temporal things and God into a common frame of reference. They never could explain how God and temporal creatures could be in any sense simultaneous so as to allow a causal relation to obtain between them. As we have seen, they eschew a consistent application of the analogy of the Special Theory of Relativity to time and eternity, which would have required that relative to God the entities/events at any time t exist at the eternal point, whereas relative to the entities/events at t God exists at t. Since they resist the idea that temporal entities/events somehow exist in or at eternity, Stump and Kretzmann failed to bring together such entities/events and God into a coherent relationship, existing as they do in two utterly disparate modes of being.

Brian Leftow's proposal is, in effect, to eliminate this defect by maintaining that temporal things really do exist in eternity, that is to say, from God's perspective all events do occur timelessly at the timeless point of eternity. Thus, God and temporal things do share a mode of existence and can be brought together within a single frame of reference, so that God and temporal entities are E-simultaneous and causally connected in that timeless state. Leftow will not allow the analogy with relativity theory to be applied symmetrically, however.[9] As a partisan of divine simplicity, Leftow cannot countenance the

---

[9] This occasions difficulty for Leftow's theory, for he affirms that while God is eternally Lord in the eternal 'reference frame', nevertheless in time he is not Lord except at the appropriate time (Leftow 1990a: 396). That is to say that God's extrinsic properties do change in the temporal 'reference frame' and, hence, God is temporal with respect to the temporal frame —and that even if relative to the eternal frame God is changeless. Thus, Leftow seems obliged to affirm with Aquinas that relative to the temporal frame, at least, God sustains no real relation to the world. The only way to prevent his solution from collapsing into the no-real-relations doctrine would therefore seem to be to deny the symmetry of the simultaneity relation implicit in relativity theory.

notion that God's mode of existence should be relative to anything, much less temporal in relation to something else.

(b) God's asymmetrical relativistic reference frame

Leftow is therefore obliged to support the asymmetrical relativity between God and temporal creatures by means of argument. Why think that relative to God temporal entities coexist with God in eternity in such a way that relative to creatures God does not coexist with them in time? Leftow bases his defence of this asymmetrical view of divine eternity on three theses:

> Zero Thesis: The distance between God and every spatial being is zero.
>
> M. There is no change of any sort involving spatial, material entities unless there is also a change of place, i.e. a motion involving some material entity.
>
> If something is temporal, it is also spatial.   (5)

On the basis of the Zero Thesis, (M), and the constriction of time to physical time, Leftow concludes that there is no change relative to God. Since an event occurring in one reference frame occurs in all (albeit simultaneous with different groups of events), explains Leftow, all events which occur in other reference frames also occur in the frame at rest relative to God. All temporal events are therefore timelessly present to God. So in answer to the question of how a timeless entity can act on events in time, Leftow asserts that 'an eternal entity acts on those temporal entities which are present with it in eternity, and these actions have consequences for temporal entities as they exist in time' (Leftow 1990a: 399). The objection under consideration fails because 'the coming to be actual in time of the events which occur at t in no way entails a change in God or in his presence to creatures. For this coming to be actual exists as well in eternity; it is just that to which God is eternally present' (ibid. 395).

The difficulty with this account of how all temporal events can be timelessly existent relative to God's 'frame of reference' is, as I have attempted to show elsewhere (Craig 1994), that there is no reason to think that there is any such 'frame of reference' because the Zero Thesis, (M), and (5) are all false. Unless some more secure foundation can be found for the existence of such a frame, it will remain problematic how all temporal events can exist timelessly relative to God.

(c) *Timeless existence of all things in eternity*

On the basis of his argument for the existence of temporal things in eternity Leftow claims that 'relative to God, the whole span of temporal events is always actually there, all at once. Thus in God's frame of reference, the correct judgment of local simultaneity is that all events are simultaneous' (Leftow 1991a: 164; 1991b: 228). This is a dark saying. If we are to make sense of it, we must construe 'always' to mean something like 'tenselessly', since God's frame of reference is timeless, not sempiternal. For the same reason, Leftow cannot mean by 'simultaneous' 'occurring at the same time', but something like 'coexistent' or 'coincident'. The statement that God judges all events to be locally simultaneous is very obscure. Leftow cannot mean that all events exist in God's timeless frame of reference, but are tenselessly ordered by a 'later than' relation such that no event occurs (tenselessly) later than any other, for that would be to affirm that there is only one time and all events occur at that moment of time. If we take literally Leftow's appeal to STR's doctrine of the relativity of simultaneity to reference frames, then we must say that just as a given set of causally unconnectable events will be calculated to sustain among themselves different relations of *earlier than, simultaneous with*, and *later than* in various reference frames, so in God's 'frame of reference' no events are judged to be earlier or later than any other or even as occurring simultaneously. Rather in God's 'frame' all events are judged to be timelessly coincident. In other words, in God's 'frame of reference' the very topology of time is voided. It would be as though one took the series of real numbers and removed from it any ordering relation such as 'greater than'. The one-dimensional temporal continuum has been divested in God's 'frame of reference' of those topological properties which make it isomorphous to a geometrical line, so that all that is left is an amorphous collection of points. Notice that in God's 'frame' even causally connected events, such as one's birth, development, decline, and death, are judged to sustain no temporal relations among themselves; they are all just timelessly coincident.

Of course, in all *physical* frames the temporal order of causally connectable events is invariant.[10] But in the special case of God, if Leftow's argument for the existence of all things in eternity is correct, this invariance does not hold with respect to his 'frame of reference'. In fact, if anyone's frame is privileged, it will surely be God's, for the relativity of simultaneity arises only for events spatially distant from the observer; judgements of local simultaneity are neither conventional nor relative. But given Leftow's Zero Thesis, all events are in a sense local for God. Therefore, his judgement that

---

[10] Padgett would invalidate Leftow's appeal to STR on this ground alone (Padgett 1993: 220–2).

all events are timelessly coincident should be absolute, and it is we who are deceived when we judge that they are temporally ordered. In fact, it is not clear to me that Leftow can avert also voiding space as well as time of any topological properties in God's 'frame of reference'. For, in relativity theory, a difference in the value of the temporal coordinate of some event relative to two distinct reference frames requires a mathematically determinate difference in the spatial coordinates of the event as well. Doubtless, Leftow would not say that the Lorentz transformation equations hold relative to God's 'frame of reference' as for physical frames. None the less, since an event's spatial coordinates are partially dependent upon its temporal coordinates, events in God's 'frame of reference', lacking any temporal coordinates, cannot be located in space either. To paraphrase Leftow: something is located in one dimension of a geometry if and only if it is located in all; so if it is correct to represent time as another dimension, it follows that whatever is not in time is not in space either: *only temporal things are spatial*. It therefore seems to follow that in God's 'frame of reference' events not only occur timelessly but spacelessly as well. The topological structure of the four-dimensional spacetime manifold has come completely unglued in the divine 'frame of reference' so that all God is confronted with is a chaotic collection of points which are ordered neither spatially nor temporally.

Leftow, however, clearly does not interpret the 'local simultaneity' of all events in God's 'frame of reference' in the above way. He states, 'In eternity events are in effect frozen in an array of positions corresponding to their ordering in various B-series' (Leftow 1991a: 170). In defending his theory against the charge that temporal beings' existing in eternity makes them eternal beings, Leftow lists the following characteristics of a temporal being in the 'reference frame' of eternity, which serve to distinguish it from an eternal being:

a. its fourth-dimensional extension or duration would have parts.
b. not all parts of its duration would occur at the same temporal present . . .
c. its duration's parts would be ordered as earlier and later.
d. in most cases, its duration would have a beginning and an end.
e. if it had no duration, still it would stand in a sequence representing the earlier–later relations obtaining between it and other events. (Leftow 1991b: 237)

These characteristics make it evident that Leftow conceives of the existence of a temporal being in eternity as the tenseless existence of its world line. In Leftow's view the entire B-series of events occurs (tenselessly) in God's eternal now. In a footnote he explains that God does not see all events spread out in one B-series, since each reference frame generates its own unique B-series. There are thus a plurality of B-series and God must be aware of all of them (Leftow 1991a: 179; cf. id. 1991b: 239; 1990a: 393).

Now this seems an eminently more reasonable account of the existence of temporal events in God's 'frame of reference', but I do not see how this account concords with Leftow's proposed theory of timeless eternity. It needs to be understood that that account does not merely eliminate the A-determinations of events (monadic predicates like *past*, *present*, and *future*) relative to the divine 'frame of reference', for STR itself takes no cognizance of such predicates in handling temporal relations among events in physical reference frames. Rather Leftow's account must also eliminate the B-determinations of events as well (dyadic predicates like *earlier than*, *simultaneous with*, and *later than*) relative to God's 'frame of reference'. For the relativity of simultaneity, which Leftow employs in order to stave off the Parmenidean conclusion that change is illusory and reality is a static whole, entails that events are classed relative to a reference frame as being either earlier than, simultaneous with, or later than any arbitrarily chosen point on the inertial trajectory of a hypothetical observer and that observers in different frames will draw at any arbitrary point on their world lines different lines of simultaneity connecting events determined to be simultaneous with that point and dividing later from earlier events. Hence, relative to God's timeless 'frame of reference', in which all events exist timelessly, God must judge of any two events that one is neither earlier than the other, nor later, nor even strictly simultaneous; they are just timelessly coexistent relative to his frame. Therefore, Leftow's theory must void even B-relations relative to the divine 'frame of reference'. Of course, an omniscient God must also know the lines of simultaneity which would be drawn by hypothetical observers relative to any physical reference frame; but in his 'frame' events are chaotically coexistent.

If the proponent of divine timelessness wants to preserve the B-relations among events, then it seems to me that his most plausible move will be to identify God's 'frame of reference' with the four-dimensional spacetime manifold itself, which God transcends, and hold that that manifold exists tenselessly. In short: the tenseless theory of time is correct. Leftow, however, denies that his theory of divine eternity entails the tenseless theory. He claims that 'a defender of God's eternity can assert that (in a strictly limited sense) one and the same event is present and actual in eternity though it is not yet or no longer present or actual in time' (Leftow 1991*a*: 165; cf. id. 1991*b*: 232). He explicates this by saying

That is, it can be true at a time t that an event dated at t+1 has not yet occurred in time, and yet also correct at t to say that that very event exists in eternity. That all events occur at once in eternity . . . does not entail that they all occur at once in time. (1991*a*)

Unfortunately, it is not apparent to me that this explication is anything but a statement of the tenseless theory. A theorist of tenseless time like Grünbaum would be adamant that at t an event at $t+1$ has not yet occurred in time (otherwise it would be earlier than t) and none the less this event exists tenselessly with as much actuality as the event at t; moreover, the tenseless theory does not assert the absurdity that all events occur at once in time, for then there would be only one moment of time! What Leftow needs to show is that his theory of the timeless existence of all things relative to God is compatible with the reality of tense, the objectivity of temporal becoming, the denial that all events exist tenselessly, and so forth.

It is at this point that the Einsteinian interpretation of STR takes centre stage in Leftow's defence. He argues,

If simultaneity and presentness are relative to reference-frames, then if present events are actual in some way in which future events are not, this sort of actuality is itself relative to reference frames. Thus there is a (strictly limited) sense in which the relativity of simultaneity entails a relativity of actuality, if one restricts full actuality to present events. (1991*a*)

This represents one way of integrating objective temporal becoming with STR, though it strikes me as enormously implausible (see Sklar 1981: 138). However that may be, Leftow explains the result of relativizing actuality to reference frames:

If we take eternity as one more frame of reference, then, we can thus say that a temporal event's being present and actual in eternity does not entail that it is present and actual at any particular time in any temporal reference frame (though it does follow that this event is, was or will be actual in all temporal reference frames). (Leftow 1991*a*: 167; 1991*b*: 234)

Again, the difficulty here is that there just does not seem to be any reference frame in which all events are present and actual, since there are in every frame spacetime regions designated absolute future or absolute past as determined by the light-cone structure at any event. The only thing corresponding to God's 'frame of reference' as described by Leftow, so far as I can see, is the spacetime manifold itself. But since it is not a reference frame, the relativity of simultaneity relation does not obtain between it and local frames. Thus, on Leftow's theory temporal becoming cannot be objective, for all events simply exist tenselessly in the four-dimensional manifold.

When pressed, Leftow shows himself prepared to fall back, if necessary, to a sort of Stump–Kretzmann model which does not appeal to the Zero Thesis, but relies exclusively on the relativity of simultaneity in order to justify the claim that actuality is reference frame dependent and therefore events which are not actual with respect to various temporal reference frames may all be actual with respect to God's 'frame' (Leftow 1990*b*: 303–21). The problem with the naked appeal to analogy, however, is that one then lacks any

justification for denying the symmetry demanded by the relativity of simultaneity, namely, that relative to temporal beings God exists in time, just as
temporal beings, relative to God, exist in eternity.

In conclusion, it seems to me that Leftow has failed to save the doctrine
of divine timelessness by means of his proposal that temporal entities exist
not only in time, but also in eternity. His argument for this asymmetrical
relation between time and eternity rests on the defective Zero Thesis, (M),
and (5). His attempt to demonstrate the compatibility of objective tense and
temporal becoming with the existence of all things in eternity never, on the
pain of incoherence, in fact advances beyond a tenseless theory of time. Leftow's
solution therefore fails to show why God's real, causal relation to the temporal world would not result in his temporal embeddedness.

5. A way out for advocates of divine timelessness?

The foregoing discussion does suggest a possible way of escape for defenders
of divine timelessness: deny the objectivity of tense and temporal becoming
and therefore also the (necessary) truth of

> If God is really related to the temporal world, God is temporal.   (4)

The argument I have given on behalf of (4) presupposes a tensed theory of
time and, hence, the reality of temporal becoming and tensed facts. But if
one embraces a tenseless theory of time, according to which there are no tensed
facts and temporal becoming is merely a subjective feature of consciousness,
then the argument is undercut. For in that case all events comprising the four-
dimensional spacetime manifold simply exist tenselessly, and God can be
conceived to exist 'outside' this manifold, spacelessly and timelessly.[11] Both
God and the four-dimensional manifold would be extrinsically timeless, in
Leftow's terminology, while the manifold is of the two alone intrinsically
temporal. But this intrinsic temporality would not involve temporal becoming
or tensed facts; all events are on an ontological par and no events are ever
objectively present. The temporality of events consists solely in their standing
in mutual relations of *earlier than*, *simultaneous with*, and *later than*, relations
which are construed as being as tenseless as the relations *less than*, *equal to*,
and *greater than*.

Thus, there is no objective matter of fact concerning any two events E and
E^* that, for example, God was once sustaining E or that God is presently

---

[11] As Denbigh puts it, 'The B-series is as if the Deity could timelessly witness all events,
laid out in order along the time coordinate, as we can witness objects laid out in space' (Denbigh
1975: 30–1). Cf. Seddon (1987: 135).

sustaining E^*. Defenders of a tenseless theory of language will argue that such tensed locutions are either reducible in meaning to tenseless expressions or have tenseless truth-conditions, thus rendering the postulation of corresponding tensed facts otiose. Moreover, since E and E^* never really come into being or pass away, no change is required on God's part to produce E at t_1 and E^* at t_2. In the absence of temporal becoming, E and E^*, as well as their respective moments, never change in their ontological status, but simply exist tenselessly (though standing in the relations *earlier/later than*). Therefore, they are changelessly real in relation to God, who also exists tenselessly. By the same act of power, God can produce tenselessly E at t_1 and E^* at t_2.[12] His single creative exercise of power is as timeless and unchanging as his intention to create, and the temporal products of his creative power exist as extrinsically timelessly as does he. The adoption of a tenseless theory of time thus gives a coherent sense to the notion that temporal events exist in eternity, and many of the statements of classical advocates of divine atemporality seem to presuppose just such a tenseless theory of time (see Craig 1985: 475–83; 1986: 93–104). Finally, since God and the creation would coexist tenselessly, then—absent any embedding hyper-time—there is no state of affairs in the actual world which consists of God's existing alone *sans* the universe, nor does the universe come into being. It begins to exist only in the sense that a metre stick has a beginning: there is a front edge to the space-time manifold, that is to say, it is finite in the *earlier than* direction. But God never brought the four-dimensional world into being; it just coexists tenselessly with him in an asymmetrical relation of ontological dependence. Had he freely determined otherwise, God could have existed alone *sans* creation; but he has freely chosen to produce a temporal world instead. Whether he chooses to exist alone or to produce tenselessly a coexisting temporal creation, God exists timelessly. Given a tenseless theory of time, God either exists tenselessly without creation or coexists tenselessly with creation, depending upon the free decree of his will, but no possible world includes both states of affairs. Thus, God, in creating the world, enters into no new relations whatsoever. He tenselessly stands in the relation of creating the big bang at t_0. The date t_0 indicates, not the time of his acting, but the time of the effect. God does not come into the relation *Creator of* with the big bang singularity at t_0 and then cease to stand in this relation to it at t_1; rather he timelessly stands in the *Creator of* relation to all events at their respective times. By a

---

[12] Objections to timeless (or even simultaneous) causation are not impressive. They are usually based on the finite velocity of physical causal influences and so find no purchase in the case of divine causal relations to the world. Le Poidevin's conclusion that causality entails time overreaches his argument, which, even if sound, would only show that a temporal cause is chronologically prior to its effect (Le Poidevin 1991: 88–94).

single, timeless act God tenselessly produces events at $t_0, t_1, t_2, \ldots$ Thus, on a tenseless theory of time, one can successfully divorce God's action from its effects in such a way that the action is timeless and the effects temporal. By denying the reality of temporal becoming and tensed facts, the advocate of divine timelessness undercuts the arguments for the necessary truth of (4), thereby allowing one to maintain God's atemporality and his creative activity in the temporal world without denying God's real relation to that world.[13]

Most advocates of divine timelessness, however, have been unwilling to pin their hopes on the tenseless theory of time. The notable exception is Paul Helm, who, more than any other contemporary philosopher or theologian, has understood the dependency of the doctrine of divine timelessness on the tenseless theory and has advocated the same. Thus, Helm brushes aside objections to divine timelessness based on God's knowledge of what is expressed by temporal indexical expressions like 'now', 'yesterday', or 'tomorrow', by citing D. H. Mellor and commenting, 'it has been plausibly argued that the use of such indexicals depends on there being a non-indexical concept of time for their proper employment' (Helm 1988: 25; cf. 78–80).[14] Helm has in mind Mellor's denial of tensed facts based on his provision of tenseless truth-conditions for tensed sentence tokens. On Helm's view the logic of temporal indexicals is no different from that of their spatial analogues, which are, of course, tenseless (ibid. 44). He explicitly advocates construing the distinction between past, present, and future as analogous to 'the spatial distinctions between here and there, and before and behind' (ibid. 47). It is clear that for Helm there is no such reality as the objective present. Without an objective present, there is no real temporal becoming, and hence, in his view, no need for any kind of ET-simultaneity such as is advocated by Stump and Kretzmann. Rather 'in creation God brings into being (timelessly) the whole temporal matrix', and 'God knows *at a glance* the whole of his temporally ordered creation' (ibid. 27, 26; cf. Helm 1993: 247). Helm is thus the one prominent advocate of divine timelessness who has advocated a coherent doctrine of God's atemporality, predicated upon the tenseless theory of time.

The conclusion to our discussion of the objection to divine timelessness based on God's action in the world must therefore be that this objection is cogent just in case a tensed theory of time is correct. If this conclusion is right, then theologians and philosophers of religion can advance the discussion of the nature of divine eternity only by tackling the difficult and multifaceted problem of the tensed versus tenseless theory of time.

[13] One of the interesting features of Padgett's treatment is that he recognizes the viability of the doctrine of divine timelessness, despite his arguments for God's temporality, if one adopts a tenseless theory, or, in his words, stasis theory, of time (Padgett 1992: 61–76).

[14] See also his interaction on this score with tenseless theorist Murray MacBeath in MacBeath and Helm (1989: 55–87).

REFERENCES

Coburn, R. (1963), 'Professor Malcolm on God', *Australasian Journal of Philosophy*, 41: 143–62.

Craig, W. (1985), 'Was Thomas Aquinas a B-Theorist of Time?', *New Scholasticism*, 59: 475–83.

—— (1986), 'St. Anselm on Divine Foreknowledge and Future Contingency', *Laval théologique et philosophique*, 42: 93–104.

—— (1994), 'The Special Theory of Relativity and Theories of Divine Eternity', *Faith and Philosophy*, 11: 19–37.

Davis, S. (1983), *Logic and the Nature of God* (Grand Rapids, Mich.: Wm. B. Eerdmans).

Denbigh, K. (1975), *An Inventive Universe* (London: Hutchinson).

Gale, R. (1991), *On the Nature and Existence of God* (Cambridge: Cambridge University Press).

Geach, P. (1969), *God and the Soul* (London: Routledge & Kegan Paul).

—— (1972), 'God's Relation to the World', in *Logic Matters* (Berkeley: University of California Press): 318–27.

Harris, J. (1992), 'God, Eternality, and the View from Nowhere', in *Logic, God and Metaphysics* (Dordrecht: Kluwer Academic Publishers): 73–86.

Hasker, W. (1989), *God, Time, and Knowledge* (Ithaca, NY: Cornell University Press).

Helm, P. (1988), *Eternal God* (Oxford: Clarendon Press).

—— (1993), 'Gale on God', *Religious Studies*, 29: 245–55.

Hill, W. (1974), 'Does the World Make a Difference to God?', *Thomist*, 38: 146–64.

Jantzen, G. (1983), 'Time and Timelessness', in A. Richardson and J. Bowden (eds.), *A New Dictionary of Christian Theology* (London: SCM): 571–4.

La Croix, R. (1982), 'Aquinas on God's Omnipresence and Timelessness', *Philosophy and Phenomenological Research*, 42: 391–9.

Leftow, B. (1990*a*), 'Aquinas on Time and Eternity', *American Catholic Philosophical Quarterly*, 64: 387–99.

—— (1990*b*), 'Time, Actuality, and Omniscience', *Religious Studies*, 26: 303–22.

—— (1991*a*), 'Eternity and Simultaneity', *Faith and Philosophy*, 8: 148–79.

—— (1991*b*), *Time and Eternity* (Ithaca, NY: Cornell University Press).

Le Poidevin, R. (1991), *Change, Cause and Contradiction* (London: Macmillan).

Lewis, D. (1984), 'Eternity Again: A Reply to Stump and Kretzmann', *International Journal for Philosophy of Religion*, 15: 73–9.

Liske, M.-T. (1993), 'Kann Gott reale Beziehungen zu den Geschöpfen haben?', *Theologie und Philosophie*, 68: 208–28.

Lucas, J. (1989), *The Future: An Essay on God, Temporality, and Truth* (Oxford: Basil Blackwell).

MacBeath, M., and Helm, P. (1989), 'Omniscience and Eternity I & II', *Aristotelian Society*, Supp. Vol. 63: 55–87.

McCann, H. (1993), 'The God beyond Time', in L. Pojman (ed.), *Philosophy of Religion* (Indianapolis: Hackett): 231–45.

Mann, W. (1983), 'Simplicity and Immutability in God', *International Philosophical Quarterly*, 23: 267–76.

Padgett, A. (1992), *God, Eternity and the Nature of Time* (New York: St Martin's Press).

—— (1993), 'Eternity and the Special Theory of Relativity', *International Philosophical Quarterly*, 33: 219–23.

Pike, N. (1970), *God and Timelessness* (New York: Schocken Books).

Seddon, K. (1987), *Time: A Philosophical Treatment* (London: Croom Helm).

Sklar, L. (1981), 'Time, Reality, and Relativity', in R. Healey (ed.), *Reduction, Time and Reality* (Cambridge: Cambridge University Press): 129–42.

Smith, Q. (1993), *Language and Time* (New York: Oxford University Press).

Stump, E., and Kretzmann, N. (1981), 'Eternity', *Journal of Philosophy*, 78: 429–58.

—— —— (1987), 'Atemporal Duration: A Reply to Fitzgerald', *Journal of Philosophy*, 84: 214–19.

—— —— (1992), 'Eternity, Awareness, and Action', *Faith and Philosophy*, 9: 463–82.

Walker, R. (1978), *Kant* (London: Routledge & Kegan Paul).

Wierenga, E. (1989), *The Nature of God: An Inquiry into Divine Attributes* (Ithaca, NY: Cornell University Press).

Wolterstorff, N. (1975), 'God Everlasting', in C. Orlebeke and L. Smedes (eds.), *God and the Good* (Grand Rapids, Mich.: Wm. B. Eerdmans): 181–203.

Wright, J. (1977), 'Divine Knowledge and Human Freedom', *Theological Studies*, 38: 450–77.

Yates, J. (1990), *The Timelessness of God* (Lanham, Md.: University Press of America).

I I

Time and Trinity

PAUL HELM

Part of the Constantinopolitanian Creed (AD 381) reads

We believe in one God, the Father All Governing, creator of heaven and earth, of all things visible and invisible; And in one Lord Jesus Christ, the only-begotten Son of God, begotten from the Father before all time.

This Creed follows the Nicene Creed, part of which reads

We believe in one God, the Father All Governing creator of all things visible and invisible; And in one Lord Jesus Christ, the Son of God, begotten of the Father as only begotten, that is, from the essence of the Father,[1]

and each Creed intends to echo the teaching of the New Testament in such passages as the first chapter of John's Gospel:

In the beginning was the Word, and the Word was with God, and the Word was God. The same was in the beginning with God. All things were made by him; and without him was not any thing made that was made.

In this chapter I wish to explore the meaning of the phrase of the Constantinopolitanian Creed 'begotten from the Father before all time' in the light of modern philosophical discussions of the relation of God to time. I shall argue, in due course, that the phrase is best understood in terms of a view which holds that God exists timelessly, though even understood in this way it is not without its difficulties. The phrase intends to express the unique relation between two persons of the Trinity; in saying that the Son is begotten, it is contrasting such a begetting with being created. Indeed according to the Creed everything that is created came into being through this eternally begotten Son. Nevertheless, the Son, not a creature, and of the same essence as

I wish to thank Robin Le Poidevin, John Taylor, and an unnamed referee for their comments on earlier drafts of this chapter.

[1] The phrase 'begotten from the Father before all time' (or 'all ages') does not occur in this, the Creed of 325, but only in that of 381. I am grateful to my colleague Dr Graham Gould for this information. The wording of the Creeds follows that of Leith (1963).

the Father, was begotten by him 'before all time'. One intended difference[2] between 'begotten' and 'created' in the Creed, I take it, is that while creation is a free activity, and what is created therefore depends on the will of the Creator and therefore exists logically contingently, what is begotten is not begotten in a parallel fashion. The Father did not volunteer the being of the Son, for how could the Son be of the same essence as the Father, be the same God as the Father, if this were so? If the Father volunteered the being of the Son, then it is possible that he might not have so volunteered, and there might not have been a Son, and this is hardly an acceptable result. For the moment I shall accept the distinction between begotten and created, but we shall be forced to scrutinize it somewhat carefully as we proceed.

And what about 'before all time'? At first sight, this is a paradoxical expression, at least if the 'before' is a temporal 'before'. For how could anything be temporally before all time? One is reminded here of Augustine's remarks in book XI of *The Confessions*, in which he defends the idea that God exists timelessly. If the expression in the Creed is not meant rhetorically, it would appear to commit its users to a similar view of God's relation to time. So the 'before' must be a non-temporal before; the sort of before used in expressions such as 'duty is before pleasure', 'age comes before beauty', and 'The Queen comes before the Prime Minister' (not in the procession, but in the Constitution).

I

We cannot proceed much further in our paraphrasing of the Creed in this fashion without meeting problems raised by those whom we shall call 'temporalists', those who maintain that God exists in time. For they will obviously understand this 'begotten before all time' somewhat differently, at least if they wish to try to make sense of this Creed in the light of their temporalism. For according to temporalism the expression 'God exists' is a tensed expression; it is an expression in the present tense, according to which 'exists' is used with the same tense as it is in 'my word-processor exists'. God exists, and yesterday he *did* exist, while tomorrow he *will* exist; again, barring an unfortunate accident, just like my word-processor. The existence of God is not like the existence of the number two, say, about which it makes no sense to raise questions about its existence yesterday, today, and tomorrow.

---

[2] To say that the Son is begotten of the Father does not necessarily exclude the idea of the Father willing the Son, and certainly not if to exclude such a thought would inevitably imply that the Father begat the Son through some imposed necessity, a necessity which would be at odds with divine impassibility.

Here is a quotation from a representative and prominent temporalist, Richard Swinburne:

God is everlasting. He determines what happens at all periods of time 'as it happens' because he exists at all periods of time. He exists now, he has existed at each period of past time, he will exist at each period of future time. This is, I believe, the view explicit or implicit in Old and New Testaments and in virtually all the writings of the Fathers of the first three centuries. (Swinburne 1994: 137–8)

Moreover, because God, according to Swinburne, exists necessarily, there was no time when he failed to exist, and there will be no time when he will fail to exist. He is backwardly and forwardly everlasting. However, his backward everlastingness is rather special. On this temporalist view God is the creator, and so there was a time when, as a result of the fiat of God, the universe came into being *ex nihilo*. This created universe, because of its law-like character, provided, in these laws, a standard of the measurement of time, a temporal metric. Before the creation of the universe God existed in time, even though when God alone existed nothing that existed changed. But the time in which God existed alone was unmetricated, unmeasurable. As Swinburne puts it:

Time . . . has a metric only if there are laws of nature, and indeed ones that attain a unique simplest form on the assumption that some periodic process measures intervals of equal time. . . . So, I claim, whether an event E_2 occurs after or before an event E_1 is independent of whether there are laws of nature; but whether there is a truth about how much later or earlier than E_2 E_1 occurs—for instance, one hour or two hours—depends on whether there are such laws of nature. (1994: 75)

Suppose a universe in which there are no laws of nature. According to Swinburne there will then be no content to talk of periods of time located by their relation to possible events; for 'the instant at which such and such would have begun or ceased to happen if initial conditions had been different' picks out no definite instant rather than any other. Even if it were true that if certain initial conditions had occurred, so-and-so would have happened (e.g. because God would have brought it about directly), there would be no truth that so-and-so would have begun/ceased to happen at this instant rather than that instant of the actual temporal continuum in which there are no laws of nature and the initial conditions did not occur (Swinburne 1994: 79).

The implication is that had the universe God created been chaotic, physically disorderly, then a metric would not have been available to him, and time, though passing, as time always does, would never be measurably passing. Some chaotic event would be later than some other chaotic event, perhaps being caused by it, but it would be impossible to say how much later. But at present we are more interested in the state of affairs before the creation of anything. Here again the implication is that time passes, but that time is unmetricated because nothing at all exists. Unlike the events that would occur in a disorderly universe, before the creation of anything nothing occurs before anything else, since nothing occurs.

Swinburne does not consider the possibility that a metric might have been provided by God for himself by means of the equivalent of an internally installed mental metronome. There seems to be no insuperable objection to the idea that God might give periodicity to time by the unfailing mental ticking of the metronome. If so, then God would not need the creation in order to generate a time-series.

More seriously, even apart from this theological fancy, it might be argued that time has an intrinsic metric.[3] Swinburne supposes that a metric depends upon lawfulness. But suppose we take a temporal series of events, E, F, and G, the product of some law. Then according to Swinburne the temporal interval between E and F must be the same as that between F and G, and a metric is born. But in virtue of what are we able to say, other than by fiat, that the temporal intervals between E, F, and G are the same? On Swinburne's view of what constitutes a metric there is no fact of the matter about whether two periods of time are the same length either before or after the creation. But if, against this conventionalism, we argued that time has an intrinsic metric, then it has that metric irrespective of whether a material universe exists.

For the moment let us leave the God envisaged by Swinburne, existing in unmetricated time before the creation of the orderly universe, and consider the well-known if not time-worn distinction in the philosophy of time, that between considering time as an A-series, and considering it as a B-series. On the A-series view, time is essentially tensed. The A-series view is the view most naturally adopted by those individuals who are in time; such individuals have a past, a present, and a future; and they habitually (if not altogether accurately) conceive of themselves as moving from the present into the future, the present obligingly moving into the past at exactly the same rate as the future moves into the present. On the A-series view, time passes; today is now, yesterday yesterday was now, and tomorrow tomorrow will be now. On the B-series view this passing of time is a myth, the myth of passage, for time, existing in a unidirectional order (as we may suppose), is tenseless. This is the view of time most naturally adopted by someone who has no point within time, or who has no interest in one point of time rather than any other point. Wednesday is before Thursday and Thursday before Friday, but though some days of the week are earlier than others (and so some

_____
[3] For recent defence of this, see Tooley (1997: section 9.7).

later than others), on the B-series view no day is ever now. Presentness, pastness, and futurity are ways we have of coping with temporal succession, not the strict truth of the matter about temporal succession. The strict truth of the matter is that the temporal order is in fact tenseless.

With this distinction in mind let us return to the God of temporalists such as that of Professor Swinburne. Unlike the atemporalist view of God, on Swinburne's view God is in a temporal series. And so it is natural to ask if the time which such a God occupies, whether we consider him before the creation of the universe, or subsequently to its creation, is most naturally thought of as an A-series, or as a B-series. Surely it is more natural to think of God as occupying an A-series view, though this does not preclude a B-series understanding. Swinburne himself would appear to give primacy to the A-series view, for he holds that there are irreducible tensed facts (Swinburne 1990).

Whether the A-series account of time is reducible to the B-series (or vice versa) is a matter of ongoing philosophical discussion. But it is hard to see how such a reduction could plausibly be effected in the case of the A-series as it features in Swinburne's view, because for Swinburne God's existence is the stopping place of the metaphysical explanation of the universe, and there are in addition good reasons for thinking that such a God exists in time. So it is a fundamental fact about reality that God is in time viewed as an A-series, with a past, a present, and a future, and that his creation is likewise in A-series time. And while it is consistent with such a position that much of our A-series talk of time can be translated into a B-series perspective, such a translation is not required, and may not be effected in the case of all such talk.

According to Swinburne the most basic form of explanation in the universe is personal explanation (1979: chs. 3 and 4). Tense is a fundamental feature of reality because it is a necessary feature of the life of the most fundamental metaphysical object. Given the basicness of the explanation of the universe in terms of the agency of God, who is in time, it does seem as if priority must be given to the A-series on Swinburne's view. This is not to say that we cannot express truths about God in B-series fashion; for a temporalist, God is earlier than the world; but Swinburne must deny that all tensed sentences about God have tenseless truth-conditions.

So according to temporalist theists such as Professor Swinburne, an ultimate feature of the world, ultimate in the sense that it is irreducible, is a process of time with God in it. But in the absence of a metric such a process will be rather peculiar, for it will be the passing of time with no possibility of the awareness of time passing. Metaphysically speaking, this will be an A-series, but it will none the less lack some of the most characteristic features of the A-series. On this A-series view of time God can distinguish between past time and future time (even though, on some views, the past no longer exists and the future does not yet exist), though not between past periods or future periods of time, for no metrication is available to him. We must conclude that in the case of a God who exists everlastingly backwards, and everlastingly forwards, and who at any time before he creates the universe cannot distinguish between a millisecond and a million years, he cannot use temporal indexicals, except in an extremely rudimentary way. The A-series thought that he will create the universe tomorrow cannot for God be a true thought, nor the B-series thought that he will create it a million years later. So what God has is an A-view of time, though not an A-*series* metricated view, since there is nothing in such a situation that will serve to establish the seriality of time. For though he can distinguish between his past and his future, God cannot take an A-*series* view of metricated time. His future is now (to God) unmetricated but metrication will emerge as soon as God wills the existence of an orderly universe.

So we may conclude that one consequence of Swinburne's temporalism, in a situation in which God exists as pure divinity, is that the most basic account of time, most basic for God, that is, is that provided by a modified and primitive A-series account. This is because of the rudimentary concept of his past and his future that, in such circumstances, God, where all explanations of contingent matters of fact stop, must possess.

II

Bearing this conclusion in mind, let us now return to the Constantinopolitanian Creed, and the phrase 'begotten from the Father before all time'. How might a Swinburnian temporalist who was sympathetic to the dogma of the Creed understand this phrase? We can best answer this question via an appreciation of how Swinburne understands the metaphysics of the Trinity. He approaches the matter by considering, in abstraction, if there could be two metaphysically necessary and necessarily good individuals (he calls them G_1 and G_2) one of which is inevitably caused by a backwardly everlasting uncaused substance (1994: 171). If the answer to this question is yes, we shall then possess the nuts and bolts of the metaphysical relationship between the Father and the Son. Swinburne argues that this relationship between G_1 and G_2 is possible, if the existence of G_2 is the result of an act of G_1's essence (1994: 173). Thus G_1 is the source of being of G_2; while G_2's metaphysical necessity is underlined by the fact that it is his (G_2's) permitting G_1 to exist which is a sufficient condition of G_1's existing.

So Swinburne presents a picture of two necessary individuals, one onto-logically necessary, the other metaphysically necessary,[4] each causing the other to exist, the one actively, the other permissively. Each, being omnipotent, has the power to annihilate the other.[5] But each of G_1 and G_2, being necessarily good, could never have a reason to annihilate the other. G_1 could never have any reason not to cause actively G_2, and G_2 never have a reason not to cause permissively G_1.

If there can be more than one divine individual, one divine individual can derive his existence from another divine individual, so long as the derivation is inevitable. For each of two divine individuals G_1 and G_2 it can be the case that there is no cause of it existing at any time while it exists, neither active cause nor permissive cause, except (directly or indirectly) an uncaused and backwardly everlasting substance, namely a divine individual, who causes his existence inevitably in virtue of his properties. If G_1, inevitably in virtue of his properties throughout some first (beginningless) period of time actively causes G_2 to exist, and thereafter permissively causes (i.e. permits) the continued existence of G_2; while G_2 is such that G_1 only exists at each period of time which has a beginning because G_2 permits G_1 to exist, then both would be metaphysically necessary—once existent, they inevitably always exist, and there is no time at which they do not exist. (Swinburne 1994: 173)

So the picture is of two divine individuals G_1 and G_2. The differences between the two are as follows. First, G_2 is inevitably derived from G_1; second, G_1 actively causes G_2 to exist throughout some first (beginningless) period of time; and thereafter permits G_2 to continue to exist; and third, G_1 only exists at each period of time which has a beginning because G_2 permits G_1 to exist. In what follows we shall be chiefly concerned with the second of these differences.

Putting this rather abstract discussion in terms of Christian doctrine, Swinburne sees the divine Father as being the source of the divine Son who inevitably exists, and who in turn inevitably (because of his nature) permits the Father to continue in existence. So the Father begat the Son in unmetric-ated time.

According to Swinburne causation is an essentially temporal notion, for he defines the past and the future in terms of causation.

So, I claim, the concepts of past and future cannot be connected to the rest of our conceptual scheme unless we understand the past as the logically contingent that is causally unaffectible, and the future as the logically contingent that is causally affectible. Unless we suppose that, any grasp we might have on the concepts would be utterly mysterious and irrelevant to anything else. (Swinburne 1994: 85–6)

---

[4] Though not ontologically necessary—see the distinction in Swinburne (1994: 146) and the critical comments on it in Alston (1997: 35–57).

[5] We may for present purposes skirt over the question of whether an omnipotent being could be annihilated; surely he would always have the power to prevent this?

It follows that neither simultaneous causation nor backward causation is logically possible. 'Any period of causing is always earlier than any effect caused, there is no simultaneous causation and no causation in a circle' (ibid. 147).

If this is so, and if $G1$ actively causes $G2$ to exist throughout some first (beginningless) period of time, then it may seem to follow that there is a period of time in which $G1$ exists and $G2$ does not exist; either that, or 'causes' is being used in a somewhat stretched or non-literal sense (stretched or non-literal, that is, when understood in terms of Swinburne's temporalist conception of causation). It is true that the period of time between $G1$'s and $G2$'s existence is not measurable, because it is a period of time in which there is no metric, and so it would be indistinguishable from either a millisecond or a million years. And it may also be temporally contiguous. For even if time intervals are unmeasurable, two events may none the less be contiguous, be such that there is no interval between them. So while there is not a temporal *interval* between $t1$ and $t2$ there is none the less a finite period of time, however short or long, in which $G1$ exists and $G2$ does not exist; a time when the Father exists but not the Son.

To this the following objection might be made, that the above argument only holds if there is a first moment of time. But if there is no first moment of time then both the Father and the Son could exist at all times, and yet the Son may depend for his existence at each of those times on the Father's prior existence. Given such a view Swinburne would be justified in maintaining that God exists in time and is forwardly and backwardly everlasting, and so God the Son could exist at all times in dependence on his Father. So if time is beginningless, this difficulty can be overcome.

This is possible, if the relation between the Father and the Son is one of ontological dependence; but if the Father causes the Son to exist, begets the Son, then this for Swinburne must be a temporal notion, even if the begetting is an inevitable consequence of the nature of the Father. And then it looks as if the onset of the begetting is not only conceptually distinct from the consummation of the begetting, but temporally distinct. Even if there was no time when the Son does not exist, perhaps because for every time the existence of the Father at that time is the cause of the existence of the Son at a slightly later time, nevertheless the Son exists at a slightly later time than the Father, and this is surely a theologically unacceptable result for Swinburne.

Let us suppose, for the moment at least, that both Father and Son are backwardly everlasting. Such backward everlastingness (coupled with Swinburne's view of time as unmetricated) does offer some anomalous if not paradoxical features which it is worth considering. Assuming an A-series understanding of God the Father's relation to time, God is able to have the following true indexical thoughts: 'Now I am begetting the Son' and 'Now I have begotten the Son.' (Or, if for every time the existence of the Father at time *t* is the cause of the existence of the Son at a slightly later time, the Father can say, 'Now I am doing *part* of what is involved in begetting the Son.' Part of, because begetting the Son involves performing an infinite number of actions at an infinite number of times. Similarly, the Father can also say, 'I will be begetting the Son, since I will be doing something that is a part of what is involved in begetting the Son.')

The begetting of the Son is thus a tensed activity, having present-tense and past-tense forms, and future continuous forms, but no future discontinuous forms; there is no true indexical thought 'I will beget the Son.' And because the Father is backwardly everlasting there is no 'then' before the 'now' of 'Now I am begetting the Son.' There is no true indexical thought of the form 'Then I had not begotten the Son' or even 'I will now beget the Son.' For the Father has always begotten the Son. The thought 'Now I have begotten the Son' indicates a completed action, but it is a completed action that was never not complete, even though it must have taken time, since for Swinburne there are no instantaneous actions.[6] Though we cannot in an unmetricated situation say how much time it took, there was no time before it, but there was time during it. (And if it takes time, it must be possible to say that it has an earlier and a later phase, even if it is not possible to say how long these phases endured).

We thus arrive at the following paradoxical features in the A-series in which God begets the Son. There is one action, the begetting of the Son, which takes time, but there was no time before it. If the action takes time, then it is possible to think of the action as having two phases, the beginning of the action, and the ending of it. No periods of time can be assigned to these phases on the basis of any facts about time; all that we can say is that the beginning phase is before, or earlier than, the ending phase, and that each phase takes some time. But since there was no time when the Son was not beginning to be begotten, there was no time when the begetting of the Son was not ending. But if there are two phases of an event, such that there is no time before each phase, then each phase must occupy the same time; so the beginning of the begetting of the Son must occupy the same time as the ending of the begetting of the Son.

Similar paradoxes can be expressed if one takes a B-series understanding of the begetting of the Son; then there is a time later than the begetting of the Son, but no time earlier than it. And, in virtue of the unmetricated nature of time, while there is a time later than the begetting of the Son, there is no time more later than that begetting; or if there is, it is impossible to tell how

---

[6] 'The present instant is the boundary between past and future' (Swinburne 1994: 81).

much more later, at least until the creation of an orderly universe. Similarly the beginning of the begetting of the Son is both earlier than and not earlier than the ending of the begetting of the Son; earlier than, because all actions, according to Swinburne, take time; but not earlier than because there is no time when the Son was not beginning to be begotten and ending being begotten.

There is, however, a second difficulty. If every causal relation is logically contingent, as Swinburne appears to hold,[7] then even if in fact there is no moment at which $G1$ exists and $G2$ does not, nevertheless it is possible that $G1$ exists and $G2$ does not exist. Swinburne appears to deny this when claiming that $G2$ is *inevitably* derived from $G1$, but it is hard to see how, if the relation between $G1$ and $G2$ is genuinely causal, there can be any logical inevitability about this derivation. Perhaps the inevitability is that of a fixed divine resolve, but however it is understood the inevitability in question cannot be that of logical or metaphysical necessity. And if the causal connection is not inevitable, then there is a possible world in which $G1$ exists and $G2$ does not exist.

Alternatively, the contingency of the causal relation may be sacrificed (perhaps on the grounds that it is an instance of the plausible principle that if X and Y are each metaphysically necessary, and X is the cause of Y, then the cause must itself be necessary). Then it is not possible for $G1$ to exist and not $G2$.

To this line of reasoning the following objections may be put:

1. *The objection from causation.* As we have noted, Swinburne regards $G1$'s active cause of $G2$, and $G2$'s permissive cause of $G1$, as each being a case of an act of nature; an act not of will, but of essence. Such acts are acts which the agent does not choose to do, but has to do in virtue of his nature. On these grounds it may be alleged that an inevitable cause is different from a non-inevitable cause in that in the case of an inevitable cause the effect is metaphysically required by the cause. That is, if there is a cause then the effect must follow as a matter of metaphysical necessity.

Let us suppose that this is correct; then the account of causation would be something like the following: A metaphysically causes B if A in virtue of his nature cannot choose not to cause B and where the existence of B temporally follows A's choice. But even if the causation is inevitable in something like the sense indicated, there would none the less still be a time at which $G2$ is caused by the nature of $G1$. This follows from the fact that for Swinburne any cause is always earlier than any effect. Of course, as we have

---

[7] 'For what causes what is logically contingent—"anything can produce anything", wrote Hume' (Swinburne 1994: 82).

already noted, his account does not require that there be any temporal *gap* between cause and effect; nevertheless, even if the effect follows the cause temporally contiguously, $G2$ exists later than $G1$.

So, to return to a point touched upon earlier, does this mean that there is a possible world in which $G1$ exists and not $G2$? If by a possible world is meant a maximally possible state of affairs then the answer must be 'Yes' if the causal relation is logically contingent; 'No' if the cause is metaphysically necessary, as according to Swinburne it is. On this view of metaphysically necessary causation there is no possible world in which $G1$ exists and $G2$ does not exist. Nevertheless $G2$ exists later than $G1$.

Swinburne also claims that $G1$ only exists because of $G2$'s permission; he says that $G2$ is such that $G1$ only exists at each period of time which has a beginning because $G2$ permits $G1$ to exist. But if we press the considerations arising from the temporality of causation the sense of 'permission' is rather curious. Since $G1$ always exists before $G2$ how can $G2$ either give (or withhold) permission for the existence of $G1$? Even if there is no time when $G1$ exists and not $G2$, because for every time t $G1$ causes $G2$ at a slightly later time, the existence of $G2$ could only permit the existence of $G1$ at a further slightly later time. $G2$ cannot retrospectively permit the existence of $G1$, and so cannot permit $G1$'s existence *simpliciter*. If we press the considerations arising from backwards everlastingness, then there is no time at which $G1$ exists and not $G2$. But 'permission' still has a peculiar sense; for how can what is metaphysically necessary be permitted? And how can what is metaphysically necessary permit an act which is an act of its nature?

2. *The objection from everlastingness.* At several points Swinburne stresses that the time in which $G1$ and $G2$ exist is everlasting:

$G2$ in turn could only exist because $G1$ everlastingly actively caused or permitted that existence. (1994: 147)

For each of two divine individuals $G1$ and $G2$, it can be the case that there is no cause of it existing at any time while it exists, neither active cause nor permissive cause, except (directly or indirectly) an uncaused and backwardly everlasting substance, namely a divine individual, who causes his existence inevitably in virtue of his properties. (Ibid. 173)

From this it is reasonable to conclude that by everlasting Swinburne means temporally ultimate. A being is everlasting if time has no beginning and if there was no time when that individual did not exist; and a cause is everlasting if time has no beginning and if there is no time when the influence of that cause was not being exerted.

If a causal relation is contingent but some causes are not temporally prior to, but simultaneous with, their effects then there is a possible world in which

$G2$ does not exist, but in any world in which $G2$ does exist there is no time in that world when $G1$ exists before it. Someone who wished to defend the Christian orthodoxy of the eternal generation of the Son by the Father might be inclined to favour such simultaneous causation. But it is a curious notion. The cases of simultaneous causation that are usually cited, such as the ball being caused to rest in the centre of a cushion by the shape and softness of the cushion, seem to be 'event-infected'. What leads us to cite these as cases of causation is that they are preceded by the event of the ball being placed on the cushion. The simultaneous causation in question is then the sustaining of the effect of that event. If some causal occurrence is not implied by the ascription of simultaneous causation then it is hard to see what justifies us in saying that it is the cushion that keeps the ball in place, and not that it is the ball that causes the cushion to be indented. As Robin Le Poidevin has put it,

if the resting ball's position brings about its effects on the cushion simultaneously, then at [*sic*] the time at which the ball was first placed on the cushion is the time at which the cushion is already compressing. (Le Poidevin 1991: 90)

Yet perhaps, in simultaneous causation, both ways of talking are justified. But this would still not satisfy the orthodox expression of the relation between the Father and the Son, for there is an asymmetrical relation between these two persons: the Son is begotten by the Father who is unbegotten, of which more below.

Reverting to our discussion of Swinburnian temporalism, our earlier conclusion is underlined. From this it appears not that there was in fact a time when the Father was and the Son was not, nor that there could have been, but that the Son was later than the Father, always later.

III

I shall close this chapter by providing a sketch of the relation between the Father and the Son if one takes an atemporal view of God. On such a view it makes no sense to think of God himself being in a temporal series, whether an A-series or a B-series. God exists atemporally, and creates the universe timelessly; the events of the created world are created as a B-series for God, and perhaps as an A-series for those agents who are in time.[8] On such a general view the Father's begetting of the Son must be a case of

---

[8] I have explored such a possibility in Helm (1997).

non-temporal causation, a relation of personal dependence. The distinction between created and begotten can in this way be maintained. In such a case, there can be no time when the Son was not.

But perhaps we are left with a difficulty of a different kind, that of giving a satisfactory account of 'begetting'. As we have seen, the Nicene thinkers wished to make a distinction between creating and begetting. The basis of this distinction, as I understand it, is that in saying that the Son is begotten of the Father the Creed denies that he is created *ex nihilo*, but rather is of the same substance as the Father, being wholly divine as he is. The problem that one is left with is with justifying the causal implications of 'begotten', and particularly the asymmetry of any such relation. For while it may be granted that 'begotten' has a meaning distinct from 'created', that meaning is not wholly distinct, in that both 'create' and 'beget' are causal notions. How can the Father beget the Son without adversely affecting the equal divinity of each and the divine unity of the pair? It would seem to follow that the Son cannot be equally divine with the Father.

Perhaps it is possible to address these questions in the following way. For an atemporalist the eternal begetting of the Son by the Father cannot express a temporal relation of any kind. The Son cannot come into being at some time after the Father, nor can he come into being at the same time as the Father. But one may be able to provide an atemporal analogue of the idea of causal simultaneity, as follows.

The idea of *temporal* simultaneity is open to a powerful objection based upon the idea of reciprocity, according to which a cause is itself affected by that upon which the cause acts in order to bring about its effect (Le Poidevin 1991: 83). But there are two reasons to suppose that such a principle, even if it is sufficient to displace the idea of full temporal simultaneity, cannot apply in the case of the Father's begetting of the Son. The first is that prior to the begetting there is no individual upon whom the begetting is wrought as a change. The begetting of the Son is not the changing of the Son, but his being brought into existence. Hence the begetter cannot be affected by that upon which he acts, since there is nothing upon which he acts. The second reinforces the first; it is that being timelessly eternal the Father is necessarily changeless, and hence for that reason alone he cannot be affected by what he changes.

The nature of the begetting must be something like the following, then: there is no state of the Father that is not a begetting of the Son, and no state of the Son which is not a being begotten by the Father, and necessarily there is no time when the Father had not begotten the Son, and no time when the Son had not been begotten by the Father. Furthermore, I think we must say that not only temporal analyses of causation do not apply in this case, neither do counterfactual analyses. One cannot sensibly state that if the Father had not begotten the Son the Son would not have existed, because the antecedent is necessarily false. There is no possible world in which the Father exists and not the Son.

The residual problem is not, how can the Son be co-divine when there was a time when the Father was and the Son was not, but, how could the Son have a timeless relation of begottenness while being equally divine with the Father? Perhaps a solution to this may be found in expunging the language of subordination entirely from the account of the Trinity, in asserting the co-equality of the Father and the Son, not their equality in every respect, but their equality in respect of divinity. The puzzle (to me at least) is why a satisfactory trinitarian doctrine may not rest content with saying that God exists in three persons co-eternal and equally divine. Is the language of begottenness and procession not a reading back into the doctrine of the Trinity *per se* of those roles which according to the New Testament each person of the Trinity adopts in order to ensure human salvation?

REFERENCES

Alston, William (1997), 'Swinburne and Christian Theology', *International Journal for Philosophy of Religion*, 41/1: 35–57.

Helm, Paul (1997), 'Eternal Creation: The Two Standpoints', in Colin Gunton (ed.), *The Doctrine of Creation* (Edinburgh: T. & T. Clark): 29–46.

Leith, John (1963) (ed.), *The Creeds of Christendom* (Garden City, NY: Doubleday).

Le Poidevin, Robin (1991), *Change, Cause and Contradiction* (London: Macmillan).

Swinburne, Richard (1979), *The Existence of God* (Oxford: Clarendon Press).

—— (1990), 'Tensed Facts', *American Philosophical Quarterly*, 27: 117–30.

—— (1994), *The Christian God* (Oxford: Clarendon Press).

Tooley, Michael (1997), *Time, Tense, and Causation* (Oxford: Clarendon Press).

Tense and Egocentricity in Fiction

GREGORY CURRIE

Fictions are said to transport us to other times and places. Not literally, of course, but in imagination. On this view, the engrossed reader of Scott's *The Talisman* imagines things the book describes as occurring in twelfth-century Palestine, not as events long ago and far away, but as *here* and *now*. As such a reader, I do not imagine those events occurring in Australia in the late twentieth century, which is where I am actually located. Rather, I imagine that my here and now is the place and time of the novel. So it is said.

If I judge that something is occurring here and now, or then and there, I locate it *egocentrically*. I locate it in relation to myself, my present time, my present place. And if I imagine, of fictional events, that their places and times are my here and now then I locate things egocentrically *in imagination*. I am interested here in the egocentric location of fictional things and the limits, if any, to such egocentric location. I am primarily interested in acts of location which involve tense, as when we say or judge that this event is now. I shall ask: are there logical or conceptual barriers to the attribution of tense to fictional events? Are there other kinds of barriers—perhaps to do with imaginative coherence—to such egocentric attributions? My answers are 'No' and 'Yes' respectively. Later on in the chapter I shall link the discussion of tense attributions in fictional contexts with some problems of temporal ordering in fictional narrative: problems about how to distinguish temporally non-standard narratives from stories involving time travel, and how it is we decide that a narrative *is* temporally non-standard.

A thread running through this discussion will be the contrast between tensed and untensed temporal predicates, or between McTaggart's A-series and his B-series. Since this distinction is well explained in Robin Le Poidevin's essay above, I shall simply assume that this contrast is understood. Coincidentally, I am going to consider an argument that Le Poidevin has constructed concerning fiction and the A-series. Some preliminaries first.

Thanks to Ian Hunt, Graham Nerlich, Greg O'Hair, and Vladimir Popescu for comments on an earlier version.

1. Distinctions

Fictions typically describe non-existents and non-occurrents: Holmes and Moriarty never existed, and the struggle at the Reichenbach Falls never took place.[1] Yet there is some inclination to say that 'Holmes and Moriarty struggled at the Reichenbach Falls' is true, especially when it is contrasted with, say, 'Holmes and Moriarty settled their differences amicably.' The best way to make sense of this inclination is to suppose that what is really true is '*It is fictional in Doyle's story* that Holmes and Moriarty struggled at the Reichenbach Falls,' while 'It is fictional in Doyle's story that Holmes and Moriarty settled their differences amicably' is false. In that case the idea that we can legitimately ascribe pastness, presentness, or futurity to fictional events should be understood as the claim that we may legitimately say, within the scope of the fiction operator, that this or that fictional event is past, present, or future. In other words, the claim under consideration is the claim that it is, or can be, legitimate for me now to say things like

> It is fictional in Doyle's story that Holmes and Moriarty struggle
> at the Reichenbach Falls in the present.   (1)

(1) can be thought of as built up from the sentence 'Holmes and Moriarty struggled at the Reichenbach Falls' by the successive application of the *presentness* operator and the *fiction* operator. Its logical form is better brought out by expressing it as 'It is fictional that (it is present that {Holmes and Moriarty struggled at the Reichenbach Falls}).' So there is an important distinction of scope to be kept in mind, and (1) is not to be confused with

> In the present it is fictional in Doyle's story that Holmes and
> Moriarty struggled at the Reichenbach Falls   (2)

in which the relations of scope are reversed. (2) I take to be uncontroversially true (true now as said by me, that is), since it has been fictional that Holmes and Moriarty struggled at the Reichenbach Falls at least since Doyle wrote the story. Anyway, I am not interested in the fortunes of (2). I am interested in statements like (1).

In (1), as I intend it to be heard, the occurrence of the fiction operator has large scope. (2), in which that operator has small scope, is rather obviously different in meaning from (1). But there *is* a way of hearing (1) in which that operator has small scope, namely

---

[1] Typically but not invariably. As we shall see later, things and events can be both real and fictional.

There is a time which is present and concerning which it is
fictional in Doyle's story that Holmes and Moriarty struggle at
the Reichenbach Falls at that time. (3)

If the present time is 1996, then (3) as uttered now is true just in case it
is fictional in Doyle's story that Holmes and Moriarty struggled at the
Reichenbach Falls in 1996. Given that the setting of the story is the late nine-
teenth century, (3) turns out to be false. But there is nothing incoherent about
(3): Doyle might have set his story in the future (his future, that is), making
it clear that 1996 was the date of the encounter. In that case (3) would be
true. (3) is certainly a candidate for truth; its truth-value simply depends on
the details of the story.

Some utterances having the grammatical form of (1) would very naturally
be heard on the model of (3). If I said that it is fictional in Doyle's story that
Holmes lived somewhere which is now occupied by a bank, I would most
charitably be understood as meaning that there is a place which is now occupied by a bank and it is fictional of that place that Holmes lived there. The
only thing controversial about this statement and about its relative (3) is that
in both cases something is said to be fictional of some real thing (a real time
or a real place). As we shall see later on, some people deny that fictions
can have real things as their subjects. However, someone might regard (3)
as acceptable while denying that attributions of tense can occur within
fictional contexts, hence denying that anything of the form of (1) could be
true. Such a person might claim that any intuitive attractiveness possessed
by (1) is accounted for by our confusing it with (3). On this view, what we
should do is reject (1) and accept (3).[2] That I deny. (3), we might say, is *de
re*; pedantically spelt out, it says that there is a time *t* such that *t* is present
and it is fictional that Holmes and Moriarty struggled at the Reichenbach
Falls at *t*. (1), as I mean it to be understood, is *de dicto*: it says that it is
fictional that there is a time *t* such that *t* is present and that Holmes and Moriarty
struggled at the Reichenbach Falls at *t*. It is the coherence of the *de dicto* (1)
which I am defending.

(2) and (3) also are rather similar; in both, the fiction operator has less
than maximal scope. But they are distinct. (3) says that there is a time *t* of
which it is fictional that something happened *at that time*; (2) says that there
is a (present) time at which it was fictional that something happened—but
not necessarily at that time.

(1) does not, strictly speaking, locate a fictional event in the A-series; it
does not say of any event that it is past, present, or future. Since the event

---

[2] I owe this point to Greg O'Hair.

referred to in (1) did not occur, no such statement could be true. (1) says that it is fictional that some event is located in the A-series, which could be true even though the event referred to in it did not occur. (Compare this with another intentional operator: 'Fred believes that E is past' does not attribute pastness to E, and is not dependent for its truth on the occurrence of E.) I shall say that statements like (1) *fictionally* locate events in the A-series. Other statements fictionally locate events in the B-series, and many of them are true. Some fictional stories are very specific about the temporal locations of their events, and very few fictions are such that one could not locate their events within some finite temporal interval. It is very natural to say, for example, that

> It is fictional in Doyle's story that Holmes and Moriarty
> struggled at the Reichenbach Falls in the late nineteenth century.　(4)

After all, the temporal location of this event is as much a part of the story as is its geographical location, and so fictionally locating events in the B-series seems to be legitimate. The only problem here is one I have already mentioned: a doubt about whether a real thing—in this case a real time interval—can be the subject of fiction. I shall address that issue briefly later on. The main question is whether there are problems about fictionally locating events in the A-series.

In summary, (1) is distinctive in various ways: in it, the occurrence of the fiction operator has large scope, thereby distinguishing it from (2) and from (3); it is *de dicto* rather than *de re*, thereby distinguishing it again from (3); it fictionally locates an event in the A-, rather than in the B-series, thereby distinguishing it from (4).

2. *Fiction and imagination*

I say that (1) is coherent, and that statements of this kind are sometimes true. I do not say that (1) is true; on the contrary, I say it is false. But my objection to (1) is not that of those who think that any statement which fictionally locates an event in the past is incoherent. Here is my view in summary:

> What is fictional in a given story is what that story makes it
> appropriate for the reader to imagine.　(A)

> Some fictions make it appropriate for the reader to imagine,
> of some event, that it is past (or present, or future).　(B)

So

> For some stories, the reader who said, 'It is fictional that this
> event is past,' would be saying something true. (C)

But

> Fictions of this kind are unusual. In 'standard' cases of fiction,
> like the Holmes stories, the story does not make it appropriate
> to imagine of any event that it is past. (D)

So

> There are barriers to the fictional attribution of tense, but the
> operation of these barriers is dependent on the genre of the
> story concerned; there are no such barriers created by 'the logic
> of fictional discourse' alone. There is nothing incoherent about
> (1); it is merely false. (E)

In this section I am concerned with propositions (A)–(C) above. The section
following (Section 3) takes a look at possible worlds analyses of fictionality.
Section 4 argues for (D) and (E).

The first issue we need to consider is what is to count as fiction. It would
not do, for example, to restrict the notion to the more conventional forms of
the novel or short story as they appear in the Western literary canon. Fiction
is a much wider notion than this; it includes, for instance, acts of story-telling
which never get embedded in writing, such as when I tell impromptu stories
to my child. My child likes to hear stories in which he himself figures, and
most of us occasionally figure in our own fantasies and daydreams. This kind
of 'audience-centred' fiction has recently become institutionalized; in so-called
interactive fictions, the reader is supposed to think of him- or herself as one
of the characters, and may be given a variety of options, the choice of any
one of which will determine that the fiction develops in one way rather than
in another. With respect to such a fiction, a reader may well think, and be
required by the narrative of the fiction itself to imagine, that some fictional
occurrence in which he himself is involved is present. For in this case the
reader is not merely imagining the occurrence of an event, but is imagining
carrying out the action which is, or brings about, that event.

Suppose that F is an interactive fiction concerning which the reader is required
at a certain point to imagine that an event, E, is happening now. How do we
get from this to the truth of

> It is fictional in F that E is present (5)

as uttered by that reader at that point? The answer is that we get there in one step by using the *Basic Thesis of Fiction*:

> What is fictional in *F* is just what *F* makes it appropriate
> for the reader of *F* to imagine.[3]   (BTF)

Fictions are, exactly, guides to imagining. In Doyle's stories, Holmes is a detective who shares his rooms and his adventures with Dr Watson, who then recounts their exploits for the public to read about. These and other things are fictional in Doyle's story; they are parts of the fiction. And they are fictional exactly because it is appropriate for the reader to imagine them. They are fictional *not* because the fiction explicitly states them. Even if it does explicitly say these things (I have not checked) that would not make them fictional, for much that is literally said in fiction is intended to be taken either in some non-literal way, or as unreliable testimony.[4] The concept of fictionality is a normative one: what is true in the fiction is what it is *correct* to imagine.

There is no use for the idea of something being fictional over and above the use to which this notion is put in determining what it is appropriate for the reader to imagine. Fictions certainly can have a content over and above what is to be imagined, as when a fiction offers some thesis about life, morality, or whatever which is offered up as a candidate for belief. But this kind of content is not *fictional* content. Sometimes it is not possible to say what is intended to be imagined, as when the outcome of the story is left open. But in that case no particular outcome is fictional. Sometimes nothing seems to be intended one way or another, as with details of the character's appearance which the novel does not describe. In that case the reader is free (within limits) to imagine whatever appearance he likes. But there again, nothing about the details of the character's appearance is fictional.

So by the BTF, if a fiction makes it appropriate for me as reader to imagine that the fictional event *E* is happening now, then I can say truly at that time, 'It is fictional that *E* is present.' And of course it will be true at that time that it is fictional that *E* is present whether I say it at that time or not, just as 'I am walking now' is true as I walk, regardless of whether I say it.

---

[3] Le Poidevin, with whom I shall be disagreeing later, agrees with this. As a reader pointed out, BTF should be relativized to times: a story which makes it appropriate to imagine that *E* is present (at the time the reader reads or otherwise detects the instruction to imagine that) may, at a later time, make it appropriate to imagine that *E* is past. Thus *F* may make it appropriate to imagine *E* is present at whatever time the reader reads the appropriate part of the text, and make it appropriate to imagine *E* is past at whatever suitably later time the reader reads the appropriate (other) part.

[4] On the concept of unreliable narration see my (1995*b*).

This is an appropriate point at which to engage with the objection that there is something incoherent in the idea of a fiction in which the reader himself is a character. This objection derives from the view that the reader is a real person and therefore not properly the subject of a fiction, and not something which can properly be referred to within the scope of the fiction operator.[5] On the face of it, this is wrong. Many events seem to be both real and fictional even in very standard forms of fiction, as the Battle of Waterloo is both real and fictional with respect to Stendhal's *The Charterhouse at Parma*; we can truly say that it is fictional in Stendhal's story that Fabrice witnessed the Battle of Waterloo. And it is surely very natural to say that the Holmes stories are set in London, not in some imagined counterpart of London. One can simply deny this, insisting that 'London', as it occurs in the Holmes stories, does not refer to London, but is rather a fictional name, like 'Sherlock Holmes'. After all, if it turns out that there is or was a real person called by that name, it would not follow that the stories are about him. Agreed. But there is a great deal of difference between 'London' and 'Holmes' as they occur in those stories. If there was a real Holmes, it was not common knowledge between readers and author that there was such a person. But it was common knowledge between Doyle and his readers that there was a city called 'London', and that common knowledge seems to inform the background and atmosphere of the stories. For this and other reasons we need not treat 'London' and 'Holmes' alike.[6] Why should we try to? Why insist that fictions cannot refer to real things? If I am right in what I have said about the nature of fiction, fictions work as instructions to get us to imagine various things. There is nothing incoherent in imagining things about yourself, as the following injunction shows: 'See that five-pound note? Imagine it is yours.'

3. A logical objection to the fictional location of events in the B-series

So far I have argued that there is a use for claims of the form of (1). The argument has been based on intuitions. But intuitions sometimes come up against hard logical facts—Cantor's set-theoretic intuitions and Russell's paradox for instance. Is there some demonstrable incoherence in claims like (1), which fictionally locate events in the A-series? Robin Le Poidevin says there is: 'placing "p is past" within the scope of the sentential operator [it is

---

[5] See Le Poidevin (1988: 252): 'We, being actual, could not be in the world of [the fiction].' See also Paul Ricœur (1985: 129): 'From the mere fact that the narrator and the leading characters are fictional, all references to real historical events are divested of their function of standing for the historical past and are set on a par with the unreal status of the other events.'

[6] On the semantics of fictional names see my (1990: ch. 4).

fictional that] will lead to an incoherent statement' (1988: 251). I can, according to Le Poidevin, say that it is fictional that p is past (or present, or future)
relative to q, where q is some other story event and the story makes it fictional
that p occurred before (or at the same time as or after) q. So I can say that
the struggle at the Reichenbach Falls is past relative to Colonel Moran's attempt
on Holmes's life, because it is part of Doyle's story that the struggle was
past at the time of that attempt.[7] But 'it is impossible to use unqualified
A-series expressions in describing a fictional or possible world becat se it is
not understood from what point of view such temporal statements are true'
(1988: 251; emphasis omitted). His argument is that the prefixing of such
an operator to a sentence locates the referents of its terms in a non-actual
possible world, 'the world of the fiction'. If I locate the fictional events in
the present, as with (1), then I am locating them in the present of *my time*,
and not the present of the world of the story. But those fictional events can
only be present with respect to the time of their own world.

Could not the time of my world and the time of the story world be the
same? Issues of transworld identification of times are relevant here. If we
can identify times across worlds, then perhaps the time of the utterer of (1)
and the time of the story can be the same. Le Poidevin does not say that such
transworld identification is impossible. He says that, if such identification is
possible, there will, in some but not all cases, be transworld identification of
times sufficient to ground claims like (1) (1988: 254). But he briefly notes
what might be a serious objection to such identification: if worlds are temporally connected we would expect them to be causally connected also, which
they are generally assumed not to be.

Now I take it that the objection, spelt out in more detail, would look like
this. Suppose we have an event e_w in world w occurring at t and an event e_u
in world u occurring at s. And suppose that t in w 'transworld identifies' with
t' in u. So t and t' are strictly identical. Assume that t' has the following one-
place property which I shall abbreviate as Φ: *any event occurring at ξ will
be causally connected with an event occurring at s*. So t has Φ. So any event
occurring at t will be causally connected with an event occurring at s. Since
e_w occurs at t and e_u occurs at s, e_w and e_u are causally connected. So w and
u are causally connected worlds. But then, contrary to assumption, they are

---

[7] In the same spirit we can account for narratorial uses of tensed expressions. Thus when
the narrator of *The Time Machine* says of the time traveller 'he may even now—if I may use
the phrase—be wandering on some plesiosaurus-haunted oolitic coral reef, or beside the lonely
saline seas of the Triassic Age' (H. G. Wells 1895: epilogue), his 'now' is defined by reference
to the time of his utterance. More specifically, it is fictional in the story that someone knows
of the activities of the time traveller and is laying out for us an account of those events, and
that at the time of the laying out the time traveller is wandering on . . . etc., etc. (This passage
from *The Time Machine* is discussed in D. K. Lewis 1976: 147.)

not worlds, for being causally disconnected from other worlds is part of what it takes to be a world.

There is a premiss to this argument which I have not so far made explicit. It is that if t has a property in world u, then t has that property in any other world in which it exists. Only then can we get from the temporal connectedness *in u* of t with s to the temporal connectedness *in w* of t with s. But this premiss is and must be denied by those who believe in transworld identity, for the premiss would lead to a straightforward contradiction in any situation where a thing is supposed to have a property in one world which it lacked in another. David Lewis has argued that there is in fact no way for the transworld identity theorists to avoid this sort of difficulty, because of the problem of what he calls 'accidental intrinsics' (see Lewis 1986). Smith has five fingers on his left hand, but it is possible for him to have six. If we take a possible worlds approach to modality and 'transworld identify' individuals, that means that there is a world where he has six. But having five (or six) fingers is an intrinsic property of Smith; it is not a relation between him and anything else. In particular, it is not a relation between him and a world. So either Smith has five fingers *simpliciter* or he does not. He cannot have both five and six. So there cannot be a world in which Smith has five and another in which that very same individual has six.

As stated, the argument only applies to what are intuitively intrinsic properties, and not to properties like Le Poidevin's Φ, which is presumably a relational property—a property which holds of t in virtue of there being a relation between t and s. But Lewis's argument could be generalized to provide the conclusion that what is intuitively a relational property supervening on a two-place relation (ξ is temporally connected to ζ) ought not to be reconstrued as a relational property supervening on a three-place relation (ξ is temporally connected in ω to ζ). In other words, the heart of Lewis's argument is an injunction against adding places; we should not construe intrinsic properties as relational properties based on two-place relations, and we should not construe relational properties based on n-place relations as relational properties based on $n+1$-place relations. At least, that is how I shall construe Lewis's argument, for only when it is so generalized can it be seen as giving aid to Le Poidevin's argument against the transworld identification of time points.

But I reject Lewis's argument. Some people used to think that velocity is an intrinsic property (perhaps some people still do). It is not, since there is the frame of reference to be considered. That was a discovery. Perhaps, if we are otherwise impressed with the claims of possible worlds semantics combined with the transworld identification of individuals, we should acknowledge another discovery: the relationality of all apparently intrinsic properties. Does that mean that we abolish the intuitive category of intrinsic properties? No. It is not as if we are saying that *having five fingers* is just like *having a fur hat*, for we are not saying that these two properties are relational to the same degree. In the old, benighted metaphysics, having five fingers was relational to degree zero, while having a fur hat was relational to degree one. In the new system everyone moves up a place; having five fingers is relational to degree one (or degree two if we bring time into the picture and wish to treat individuals as persisting through time) while having a fur hat is relational to degree two (or three). We still have all the distinctions we had before (see Jackson 1992).

So I think we can answer Le Poidevin's objection to the transworld identification of time points. And in answering it we are not making special, ad hoc provision to cope with this difficulty; the problem is, as Lewis has shown, a perfectly general one which concerns a whole range of properties which a whole range of entities can have. Still, I do not want to give the impression that in order to make sense of claims like (1) we have to believe in the transworld identification of time points. We might instead opt for Lewis's own strategy of finding counterparts in other worlds for this-worldly times (Lewis 1986). One way to do that is suggested by Le Poidevin himself, though he seems to be thinking of this as a strategy for the transworld identification of times (Le Poidevin 1988: 253). The suggestion is that we identify the counterparts of a time by reference to events located at that time. Thus if, in world *w*, all the events in the set $\{e_1, \ldots e_n\}$ occur at *t*, then the occurrence of some suitable subset of this set in world *u* at *t′* would be sufficient for us to say that *t* and *t′* are counterparts.

Counterpartness is not identity. But we should not conclude from this that a counterpart theorist must reject a claim like (1). For Lewis's theory of counterparts is not intended as a revisionary metaphysics which would call into question intuitive judgements like 'If I had chosen the other box I would have won the money.' While it is strictly speaking incorrect on a counterpart-theoretic approach to suppose that I am considering numerically the same individual (myself) in different circumstances when I make such a judgement, counterpart theory is intended to provide a way of reading such counterfactual claims under which they come out true. If counterpart theory turned out to have the consequence that on no plausible reading did such a claim come out true, then that alone would seem to be reason for rejecting it.[8] Similarly for theories of fiction. An approach to fiction via counterpart theory under which all formulations of the idea that fictions can be set at real times and places come out false would be unacceptable. Either a counterpart-theoretic approach

---

[8] See Kripke (1980: 195–8), and the response from Lewis in his (1986).

must validate that idea or the approach must be rejected; it cannot be used as grounds for refuting the idea. For whatever we decide in general about the relation between the times of different worlds—be it straightforward stip- ulative identity, identification via qualitative similarity, or counterpartness —we should not lose sight of a very basic intuition about the time of the Holmes stories and the real time of the actual world: they are the same.[9] Just as readers of the story are supposed to imagine that the real London is the setting of the stories, so they are to imagine that a certain vaguely specified stretch of real time is the time during which the story unfolds. And so it is for just about any conventional fiction; the time of *The Talisman* is the real twelfth century, the time of *The Great Gatsby* is the real 1920s, and so on. An analysis of fictionality—in possible worlds terms or any others—which offers no basis for such judgements as these is scarcely credible.

So the adoption of a possible worlds approach to fiction, combined with a theory either of transworld identity or of counterpartness, does not auto- matically create problems for the integrity of claims like (1). But there are yet other options. The possible worlds approach to fictionality is not the only viable, or even the best, approach there is. The ubiquitous phrase 'the world of the fiction' can easily give the impression that there is simply no alter- native to thinking about fiction in terms of possible worlds. But on any con- sidered account this is a very misleading phrase, and no respectable theory can endorse the notion that there is such a thing as *the* world of the fiction, that fictions are paired uniquely with individual worlds. Certainly Lewis's theory does not, for on Lewis's theory each fiction is associated with a class of worlds, the members of which need to be compared according to their similarity to the actual world, or to worlds which conform to how the actual world is believed to be, in order for us to make a judgement about what is fictional in a given story.[10] Lewis's theory, whatever its merits, gets no sup- port from, and gives no support to, the intuitive but flawed notion of 'the world of the fiction'.[11]

Before I briefly sketch a theory of fictionality which does not make use of possible worlds semantics, I will say something about what a theory of

---

[9] Robert Paul Wolff has claimed that 'to say of an event that it is real . . . is no more and no less than to assign it to an objective time location' (1990: 212). Conversely, failure of location in objective time is the mark of the imaginary: 'Lord of the Rings is shown to be imaginary by the impossibility of establishing any temporal translation from its elaborate chronology, to the unreal chronology of the real world' (p. 213). But *Lord of the Rings* is an exceptional fiction, and the failure of its time to connect with ours is incidental to its status as fiction. For the Holmes stories and many other cases of the fictional do so connect: their time is the real time. Of course we say 'Holmes flourished in the 1890s' without asserting it. But nor do we assert when we say 'Holmes was a detective.' Location in time is no more significant for the real/fictional distinction than any other property of things or events.

[10] See Lewis (1978).   [11] For criticism of Lewis's theory, see my (1990: ch. 2).

fictionality is intended to do. We already have, in a sense, a theory of fiction-ality: the Basic Thesis of Fiction, according to which what is fictional in a given story is what it is appropriate for the reader to imagine. But this thesis does not shed any light on the epistemological issues. How is it that readers decide what is fictional, and hence decide what it is appropriate to imagine? Lewis's possible worlds approach offers an answer to this, by postulating that readers make judgements based on their estimates of similar-ity between worlds in which the text of the story is uttered as known fact and the actual world (or, in a more sophisticated variant which Lewis offers, their estimates of similarity between worlds in which the text of the story is uttered as known fact and worlds which conform to how the actual world is or was believed to be by the author's community).[12] My own answer to this question is quite different. I say that readers work out what is fictional by working out, on the basis of their reading of the text together with certain background assumptions, what it is reasonable to believe the author intended the reader to imagine. I have elaborated this thesis elsewhere and I will not repeat myself here.[13] I mention it to indicate the existence of an account of fictionality which appeals not to possible worlds but to intention. And on this account there is no difficulty in principle in accepting claims like (1). If what is fictional is what it is reasonable to believe was intended to be imagined, then anything which could reasonably be thought of as intended to be im-agined would be a candidate for fictionality. And there is no barrier to the idea that someone can intend me to imagine that something is past, present, or future.

Where are we? We have every reason to accept and no reason, logical or otherwise, to reject the idea that a fictional time can also be a real time, so we have no reason to reject the idea that a reader can think of a fictional time as past. So if there are narratives which make it appropriate for readers to imagine, at a given stage of their reading, that a time is past, those narrat-ives make it the case, at that stage, that it is fictional that this time is past.

4. Egocentric location and imagination

So I say that there is nothing incoherent in the idea of imaginatively attribut-ing pastness, presentness, or futurity to fictional events, and hence nothing incoherent in the idea of the fictional location of events in the A-series. Such attributions are examples of what I have called egocentric locations of the

---

[12] See Lewis (1978: analyses 2 and 3). Lewis would not, I think, want to be read as claiming that readers are always or usually conscious of making these inferences.

[13] See my (1993). The formulation in that article improves on the very rough one given here.

fictional. Other examples are: imagining Holmes and Moriarty struggling *here* (wherever I happen now to be); imagining *myself* struggling with them; imagining *seeing* them struggle (here and now).

But while all these notions are logically coherent, there are sound reasons for thinking that certain kinds of egocentric location of the fictional threaten narrative coherence and a proper imaginative response to the fiction concerned. Some writers on fiction have taken the contrary view and have argued that the appropriate imaginative response to standard fictional forms *is* to locate its characters and events egocentrically. Thus A. A. Mendilow: 'The reader if he is engrossed in his reading translates all that happens from this moment of [fictional] time onward into an imaginative present of his own and yields to the illusion that he is himself participating in the action or situation.'[14] Mendilow says that the reader 'yields to an illusion' that he is a participant. Taken literally, that is hardly credible; readers of fiction cannot reasonably be thought of as deluded into thinking that they are at Waterloo or beside Anna Karenina as she prepares to jump under the train. Perhaps he means that they *imagine* (very vividly imagine) that they are. Even that is implausible. What of those events of which it is fictional that they occur when no one is there, or at least no one other than the explicitly described characters? Are we to imagine that those events are witnessed only by Bert and Ernie, if that is what is fictional, and also that we ourselves witness them? That would be to have contradictory imaginings. From where and in what guise are we to imagine ourselves witnessing them, if it is fictional that the other characters do not see us or notice us in any way? Unless fictions are explicitly designed with 'holes' in which the reader is intended to insert himself or herself, imagining the fiction here and now quickly collides with the fictional content of the story.

The idea that we do imaginatively locate the fiction in the here and now has most often been argued for in connection with film.[15] The nature of the filmic medium makes the idea in some ways an attractive one. There we sit, engrossed by visual images of people doing things. It seems that we look directly at the characters, if not in reality at least in imagination. Indeed, the intrusion of the camera into the space of the action seems to ground our own imagined presence in that space; we imagine seeing these events from the point of view of the camera. And of course we imagine seeing them now; we imagine seeing them at the time we really see the screen images.

But this view is heir to a number of difficulties, some of which I have already mentioned. What are we to imagine when the film makes it plain that

[14] Mendilow (1972: 96–7); see Sternberg (1978: 21–2) for criticism.
[15] See e.g. Lotman (1976: 77): 'In every art which employs vision and iconic signs there is only one possible artistic time—the present.' See also Balázs (1970: 120).

the events we are supposedly witnessing are unseen by anyone? It does not help to suggest, as people occasionally do, that we are to imagine watching 'from outside the world of the fiction'. In most films, the possibility that the events of the story could literally be *seen* from another world is ruled out. Imagining that we see in an extramundane fashion would be more, not less, in conflict with the fiction.

I suggest then that we would be better off if we could fashion an account of our imaginative engagement with fiction which did not have as a consequence the idea that we are, in imagination, present at the action. Such an account will have it that the mode of imagining which is standard for our encounters with regular genres of fictions is *impersonal* imagining: we imagine the occurrence of the events of the story, but we do not imagine them egocentrically. Such an account will not rule out the possibility of our imagining fictional events as, say, present. We need, after all, to make room for interactive fictions and the stories I tell my child. But it should make generous room for those fictions—standard forms of the novel and the film, for example—which do not ask us to imagine things of ourselves.

I am not here going to elaborate any specific version of such an account. There are a number of problems to be overcome if any such an account is to be viable, and I shall briefly discuss two of them which impinge directly on the issue of tense.[16] I shall discuss them for the case of film which, for technical reasons, is a medium better suited to the exploration of the representation of time,[17] but the discussion extends easily to cover other forms like the novel.[18]

If our imaginings are to be impersonal, how can they accommodate our intuitions about *anachrony*, the reordering of story-time by narrative?[19] The distinctive modes of anachrony are the flashback and the flashforwards (or the analepsis and the prolepsis). But to see a part of the narrative as involving a flashback is surely to see it as a movement from the *present* of the story into the past, and similarly with the flashforward. Thus anachrony seems to require explication in terms of the A-series. We might try to define anachrony in B-series terms alone, thus:

[16] For a more general discussion see my (1995*a*: chs. 5 and 6).

[17] Some relevant differences between film and literature are discussed in the appendix to ch. 7 of my (1995*a*).

[18] This idea has implications in other areas which I cannot properly explore here. For example, there used to be talk in aesthetic circles of 'psychical distancing', though a satisfactory account of what this might involve never seemed to arrive. I think there is something to the idea that the proper appreciation of art, and in particular of narrative art, requires some sort of distancing. I suggest that we can understand this in terms of the kind of imagining appropriate to fiction. An appropriately distanced engagement with the fiction is one in which the imaginings are impersonal.

[19] See Gérard Genette (1980).

Film *F* contains anachrony iff *F* contains representations of
fictional events *X* and *Y*, where the representation of *X* in
viewing time is after that of *Y*, but it is fictional that the time
of the occurrence of *X* is before that of *Y*.   (A)

But this raises problems of various kinds.[20] One, put to me by David Lewis,
concerns the relation between anachrony in cinematic fiction and those cinematic narratives which involve time travel. In a time-travel story the narrative might begin in, say, 1984, during which plans are made for a journey into
the past; later in the narrative we are presented with events which take place
at the time travelled back to, say 1954. It is intuitively wrong to assimilate
cinematic (or other) narratives involving time travel to anachronous narratives;
there seems to be a sense in which, in the time-travel story I described, the
events of the story are presented in the chronologically *correct* sequence; after
all, the journey back to 1954 takes place *after* the events in 1984, which are
presented first. How, otherwise, would the events of 1984 constitute a preparation for the journey? But then the objection to my definition (A) is, exactly,
that it conflates anachronous narratives and narratives involving time travel.[21]
For it is true, concerning the time-travel story just described, that events occurring in 1954 are shown later in viewing time than events occurring in 1984.

Lewis was good enough to suggest a way out of the difficulty for me. The
objection shows not that there is an error in definition (A), but rather that
there is an ambiguity in its statement. That is, there is an ambiguity in the
expression 'the time of the occurrence of *X*'. Is this supposed to refer to objective time, or to what is sometimes called personal or subjective time? With
time travel, as it occurs in stories and as it might occur in reality if it ever
does, there is a disparity between objective and personal time. The traveller
travels back to a time earlier in objective time than the time she left; from
1984 to 1954, as it might be. But for her, the events she encounters in 1954
are later (say, an hour later) than the events she previously encountered in
1984. Here the time traveller's journey is thirty years into the objective past,
and one hour into her personal future.[22]

Normally, in stories and in reality, objective time and personal time run in
the same direction and at the same rate, and there is no need to distinguish

---

[20] The following four paragraphs follow closely the exposition in my (1995*a*: section 7.2).

[21] It is possible for there to be a time-travel story that would not count as anachronous according to definition (A). That would be a narrative in which the events 'travelled back to' in
1954 are presented first in viewing time and the events leading up to the journey in 1984 are
presented later in viewing time. So the objection is not that (A) makes all time-travel stories
come out as anachronous, but rather that it makes some of them so appear—and in fact all the
time-travel narratives I know about would come out as anachronous on the definition.

[22] On the distinction between objective and personal time see e.g. David Lewis (1976).

between them. Time travel occurs when they come apart. How is that possible? Perhaps in this way: that the direction of time is the direction of causation —the direction from causes to effects. That is why we can remember the past but not the future, and why, more generally, we are familiar with traces of the past in the present, but never encounter traces of the future in the present (unconfirmed reports of premonitions aside). But suppose that not all causal processes move in the same direction, that there is a small minority of causal processes that swim against the tide. In that case we could say that the *predominant* direction of causation is the direction of objective time, and that, given this direction, objects undergoing reversed causation are travelling back in objective time. But for those involved, if they are sentient creatures capable of thought and memory, their journey backwards will end after it began; for them, the reversing of causal processes will mean that objectively later states of consciousness will affect objectively earlier states, and the travellers will remember doing things in 1984 when they get to 1954; if the journey takes a significant amount of time, they may end the journey hungrier than they began it, have fuller beards and longer fingernails. The journey ends later in their personal time, and earlier in objective time.

With this distinction between objective and personal time, we can solve our problem. In a time-travel story, events that occur earlier in objective time may be recounted after events that occur later in objective time. But according to the model of time travel just proposed, the events occurring earlier in objective time are occurring later in personal time. So if we take 'the time of the occurrence of X', as that expression occurs in (A), to refer to a character's personal time, time-travel stories will not count as anachronous according to that definition. Of course it would be awkward to interpret 'the time of the occurrence of X' in (A) as sometimes referring to objective time and sometimes to personal time, according to whether the work in question is a time-travel story or not. But we need not do that. We may simply say that 'the time of the occurrence of X' always refers to personal time, which, in the case of a story which involves no time travel (whether it involves anachrony or not), will automatically coincide with objective time.

Another difficulty for the idea that the standard mode of imagining for fiction is impersonal imagining concerns the problem we have in making temporal sense of some complex fictional narratives. While writers on fiction have often discussed particular cases of this phenomenon, there has been little systematic work done on providing a theoretical framework for understanding the process by which we make judgements about the temporal structure of a story. But an interesting article by I. S. Talib makes a start in this direction, in the process of considering the notorious difficulties involved in understanding the narrative of *Nostromo* by Joseph Conrad.[23] Anachronous narratives are understood, Talib suggests, by the reader locating a narrative 'now' with respect to which other events of the story can be understood as past or future. In fact, if I understand Talib rightly, there is to be a hierarchy of nows, for every story event is now from the point of view of someone temporally located at its occurrence. The key step in understanding temporally complex narratives like *Nostromo* is to locate a highest now, the now with respect to which all other story events count as past.

While some of the details of Talib's account seem to me wrong—in particular the idea that the *latest* events of the story are maximally privileged as now—the idea that we judge the to-ing and fro-ing of narrative time by anchoring the narrative to a sequence of events considered to be present is a very appealing one. But such an idea has no place in the theoretical structure I have announced, because I do not allow that there is, in standard cases, a set of fictional events which the audience think of as present. It looks as if my denial that the audience regard any fictional events in conventional fictional works as present leaves us unable to explain the judgements we make about the to-ing and fro-ing of narrative time.

I answer the objection in this way. I agree with Talib that temporally complex narrative is understood by locating a sequence of events as privileged. But I deny that this must involve our coming to regard the privileged events as present. There are ways to privilege a certain sequence of story events, without thinking of them as present or as past in relation to me. An analogy might help here. In understanding a complex technical drawing, it might help if I can locate a certain subset of the lines of the drawing as horizontal, and think of the other lines as deviating in various ways from that orientation. But that need not mean that I think of the set of horizontals as parallel to *my* horizon; the drawing itself might be placed in such a way that the lines are not parallel to my horizon. I privilege the lines so as to make sense of the drawing, but I do not privilege them egocentrically.

How, then, do we privilege a certain sequence of story events as the fixed point from which the flashback and the flashforward are measured? Strictly speaking this is a question for empirical psychology, and a somewhat under-investigated one. But perhaps a philosopher may make a few suggestions that someone else will test. As in other matters, we probably operate with a set of default assumptions, and a set of loose criteria for deviating from them. Thus we assume that *all* the fictional events are temporally standard unless the assumption clashes with the demand of narrative coherence. If we cannot make a coherent story on the assumption that the narrative begins at the

---

[23] Talib (1990: 1–21).

beginning and goes on till the end, we look for ways in which some of the material can be thought of as 'out of temporal sequence'. Which material, and in which direction out of sequence (flashback or flashforward?), probably depends on a number of factors like simplicity and the presence of significant relations, particularly of a psychological kind, between the story events. Thus we tend to avoid, where we can, treating massive chunks of the narrative as out of temporal order, preferring to *minimize* the deviation from what is temporally standard. Sometimes we are presented with fictional events A and B, where event B seems to have occurred prior to event A, though B is shown after A. Should we identify A as standard and B as a flashback, or A as a flashforward and B as temporally standard? There are cases where the one description seems as good as the other, and we feel we are losing our grip on what is temporally standard and what is not. But sometimes the decision seems intuitive and non-arbitrary. That might be because of the psychological connections between A and B. Is someone at A remembering B? Is a character at B having a premonition of A (a less common occurrence)? In the former case the description of B as a flashback seems natural, and in the latter case the description of A as a flashforward does. None of these decisions seems to involve the necessity that we think of what is temporally standard as present.[24]

5. *Conclusions*

I have argued that the relations between tense and fiction are rather complicated. There are, I claim, no logical or conceptual barriers to the idea that we locate ourselves egocentrically in relation to fiction. But for particular fictions—indeed, for most fictions of any complexity and interest—such acts of egocentric location would be out of step with the project of engaging imaginatively with the work. While it is at first sight attractive to suppose that we imagine that those fictional events are present (temporally and spatially), I believe we can reconstruct our most robust intuitions about fictional narratives and our relations to them without this assumption. And I have tried to show this for the case of some intuitions we have concerning anachrony in fictional narrative.

REFERENCES

Balázs, B. (1970), *Theory of the Film* (New York: Dover Publications).
Currie, G. (1990), *The Nature of Fiction* (Cambridge: Cambridge University Press).

[24] See my (1995*a*: ch. 6).

—— (1993), 'Interpretation and Objectivity', *Mind*, 102: 413–28.

—— (1995*a*), *Image and Mind* (Cambridge: Cambridge University Press).

—— (1995*b*), 'Unreliability Refigured', *Journal of Aesthetics and Arts Criticism*, 53: 19–29.

Genette, G. (1980), *Narrative Discourse: An Essay in Method*, trans. J. E. Lewin (Ithaca, NY: Cornell University Press).

Jackson, F. (1992), 'Metaphysics by Possible Cases', in Brian Garrett and Peter Menzies (eds.), *1992 ANU Metaphysics Conference*, Working Papers in Philosophy 2 (Canberra: Research School of Social Sciences, Australian National University).

Kripke, S. (1980), *Naming and Necessity* (Oxford: Basil Blackwell).

Le Poidevin, R. (1988), 'Time and Truth in Fiction', *British Journal of Aesthetics*, 28: 248–58.

Lewis, D. K. (1976), 'The Paradoxes of Time Travel', *American Philosophical Quarterly*, 13: 145–52; repr. in *Philosophical Papers*, vol. ii (New York: Oxford University Press, 1986).

—— (1978), 'Truth in Fiction', *American Philosophical Quarterly*, 15: 37–46; repr. in *Philosophical Papers*, vol. i (New York: Oxford University Press, 1983).

—— (1986), *On the Plurality of Worlds* (Oxford: Basil Blackwell).

Lotman, J. (1976), *Semiotics of the Cinema*, trans. Mark E. Suino, Michigan Slavic Contributions 5 (Ann Arbor: University of Michigan Press).

Mendilow, E. E. (1972), *Time and the Novel* (New York: Humanities Press).

Ricœur, P. (1985), *Time and Narrative*, vol. iii (Chicago: Chicago University Press).

Sternberg, M. (1978), *Expositional Modes and Temporal Ordering in Fiction* (Baltimore: Johns Hopkins University Press).

Talib, I. S. (1990), 'Conrad's Nostromo and the Reader's Understanding of Anachronic Narratives', *Journal of Narrative Technique*, 20: 1–21.

Wells, H. G. (1895), *The Time Machine, an Invention* (London: Heinemann).

Wolff, R. P. (1990), 'Narrative Time: The Inherently Perspectival Structure of the Human World', in P. French, T. Uehling, and H. Wettstein (eds.), *Midwest Studies in Philosophy*, xv: *The Philosophy of the Human Sciences* (Notre Dame, Ind.: University of Notre Dame Press).

BIBLIOGRAPHY

The following is a select bibliography of items, published from 1980 onwards, either specifically on the debate between the tensed and tenseless theories, or on closely related topics. The original date and place of publication are given. Some of the items have been reprinted in Oaklander and Smith (1994), and these are marked with an asterisk.

ARTHUR, R. T. W. (1982), 'Exacting a Philosophy of Becoming from Modern Physics', *Pacific Philosophical Quarterly*, 63: 101–10.

BEER, MICHELLE (1988), 'Temporal Indexicals and the Passage of Time', *Philosophical Quarterly*, 38: 158–64.*

BERNSTEIN, MARK H. (1992), *Fatalism* (Lincoln: University of Nebraska Press).

BIGELOW, JOHN (1991), 'Worlds Enough for Time', *Noûs*, 25: 1–19.

BRANDON, E. P. (1986), 'What's Become of Becoming?', *Philosophia*, 16: 71–7.

BULLER, DAVID J., and FOSTER, THOMAS R. (1992), 'The New Paradox of Temporal Transience', *Philosophical Quarterly*, 41: 357–66.

BUTTERFIELD, JEREMY (1984a), 'Prior's Conception of Time', *Proceedings of the Aristotelian Society*, 84: 193–209.

—— (1984b), 'Dummett on Temporal Operators', *Philosophical Quarterly*, 34: 31–42.

—— (1985a), 'Spatial and Temporal Parts', *Philosophical Quarterly*, 35: 32–44.

—— (1985b), 'Indexicals and Tense', in Ian Hacking (ed.), *Exercises in Analysis: Essays by Students of Casimir Lewy* (Cambridge: Cambridge University Press): 69–87.

—— and STIRLING, COLIN (1987), 'Predicate Modifiers in Tense Logic', *Logique et analyse*, 117–18: 31–50.

CAMPBELL, JOHN (1994), *Past, Space and Self* (Cambridge, Mass.: MIT Press).

CARGILE, JAMES (1989), 'Tense and Existence', in John Heil (ed.), *Cause, Mind and Reality: Essays Honoring C. B. Martin* (Norwell: Kluwer).

CHAPMAN, T. (1982), *Time: A Philosophical Analysis* (Dordrecht: D. Reidel).

CHISHOLM, RODERICK (1990), 'Referring to Things That No Longer Exist', in James E. Tomberlin (ed.), *Philosophical Perspectives*, vol. iv (Atascadero, Calif.: Ridgeview): 545–56.

CHRISTENSEN, FERREL (1993), *Space-Like Time* (Toronto: University of Toronto Press).

COCKBURN, DAVID (1987), 'The Problem of the Past', *Philosophical Quarterly*, 37: 54–77.

—— (1997), *Other Times: Philosophical Perspectives on Past, Present and Future* (Cambridge: Cambridge University Press).

CRAIG, WILLIAM LANE (1985), 'Was Thomas Aquinas a B-Theorist of Time?', *New Scholasticism*, 59: 475–83.

CRAIG, WILLIAM LANE (1990*a*), 'Aquinas on God's Knowledge of Future Contingents', *Thomist*, 54: 33–79.

—— (1990*b*), 'God and Real Time', *Religious Studies*: 335–59.

—— (1996*a*), 'Tense and the New B-Theory of Language', *Philosophy*, 71: 5–26.

—— (1996*b*), 'The New B-Theory's *Tu Quoque* Argument', *Synthese*, 107: 249–69.

CRESSWELL, M. J. (1990), 'Modality and Mellor's McTaggart', *Studia Logica*, 49: 163–70.

CURRIE, GREGORY (1992), 'McTaggart at the Movies', *Philosophy*, 67: 343–55.

DENYER, NICHOLAS (1981), *Time, Action and Necessity* (London: Duckworth).

DIEKS, DENNIS (1988), 'Special Relativity and the Flow of Time', *Philosophy of Science*, 55: 456–60.

DORATO, MAURO (1995), *Time and Reality: Spacetime Physics and the Objectivity of Temporal Becoming* (Bologna: Cooperativa Libraria Universitaria Editrice Bologna).

EVANS, GARETH (1985), 'Does Tense Logic Rest on a Mistake?', in *Collected Papers* (Oxford: Clarendon Press): 343–63.

FARMER, DAVID J. (1990), *Being in Time: The Nature of Time in Light of McTaggart's Paradox* (Lanham, Md.: University Press of America).

FORBES, GRAEME (1987), 'Is There a Problem about Persistence?', *Aristotelian Society*, Supp. Vol. 61: 137–55.

GARRETT, B. (1988), ' "Thank Goodness That's Over". Revisited', *Philosophical Quarterly*, 38: 201–5.*

HACKER, PETER (1982), 'Events and Objects in Space and Time', *Mind*, 92: 1–19.

HASLANGER, SALLY (1989), 'Persistence, Change and Explanation', *Philosophical Studies*, 56: 1–28.

HELM, PAUL (1988), *Eternal God: A Study of God without Time* (Oxford: Clarendon Press).

—— (1989), 'Omniscience and Eternity', *Aristotelian Society*, Supp. Vol. 63: 75–87.

HESTEVOLD, H. SCOTT (1990), 'Passage and the Presence of Experience', *Philosophy and Phenomenological Research*, 50: 537–52.*

HIGGINBOTHAM, JAMES (1995), 'Tensed Thoughts', *Mind and Language*, 10: 226–49.

HORNSTEIN, NORBERT (1990), *As Time Goes By: Tense and Universal Grammar* (Cambridge, Mass.: MIT Press).

HORWICH, PAUL (1987), *Asymmetries in Time* (Cambridge, Mass.: MIT Press).

JOHNSTON, MARK (1987), 'Is There a Problem about Persistence?', *Aristotelian Society*, Supp. Vol. 61: 107–35.

KAPLAN, DAVID (1989), 'Demonstratives: An Essay on the Semantics, Metaphysics and Epistemology of Demonstratives and Other Indexicals', in Joseph Almog et al. (eds.), *Themes from Kaplan* (Oxford: Oxford University Press): 481–563.* (in part)

KIERNAN-LEWIS, J. D. (1991), 'Not Over Yet: Prior's "Thank Goodness" Argument', *Philosophy*, 66: 242–3.*

—— (1994), 'The Rediscovery of Tense: A Reply to Oaklander', *Philosophy*, 69: 231–3.

KUHN, STEVE T. (1989), 'Tense and Time', in D. Gabbay and F. Guenthner (eds.), *Handbook of Philosophical Logic*, iv: *Topics in the Philosophy of Language* (Dordrecht: D. Reidel): 513–52.

LEFTOW, BRIAN (1991), *Time and Eternity* (Ithaca, NY: Cornell University Press).

LE POIDEVIN, ROBIN (1988), 'Time and Truth in Fiction', *British Journal of Aesthetics*, 28: 248–58.

—— (1991), *Change, Cause and Contradiction* (London: Macmillan).

—— (1993), 'Lowe on McTaggart', *Mind*, 102: 163–70.

—— (1996), 'Time, Tense and Topology', *Philosophical Quarterly*, 46: 467–81.

—— and MACBEATH, MURRAY (1993) (eds.), *The Philosophy of Time* (Oxford: Oxford University Press).

—— and MELLOR, D. H. (1987), 'Time, Change and the "Indexical Fallacy"', *Mind*, 96: 534–8.

LEVISON, ARNOLD B. (1987), 'Events and Time's Flow', *Mind*, 96: 341–52.

LEWIS, DELMAS (1988), 'Eternity, Time and Tenselessness', *Faith and Philosophy*, 5: 72–86.

LOIZOU, ANDROS (1986), *The Reality of Time* (Aldershot: Gower Publishing Co.).

—— (1996), *Time, Embodiment and the Self* (Aldershot: Avebury).

LOWE, E. J. (1986), 'On a Supposed Temporal/Modal Parallel', *Analysis*, 46: 195–7.

—— (1987*a*), 'The Indexical Fallacy in McTaggart's Proof of the Unreality of Time', *Mind*, 96: 62–70.

—— (1987*b*), 'Lewis on Perdurance versus Endurance', *Analysis*, 47: 152–4.

—— (1987*c*), 'Reply to Le Poidevin and Mellor', *Mind*, 96: 539–42.

—— (1988), 'Substance, Identity and Time', *Aristotelian Society*, Supp. Vol. 62: 61–78.

—— (1992), 'McTaggart's Paradox Revisited', *Mind*, 101: 323–6.

—— (1993), 'Comment on Le Poidevin', *Mind*, 102: 171–3.

LUCAS, J. R. (1989), *The Future: An Essay on God, Temporality and Truth* (Oxford: Basil Blackwell).

MACBEATH, MURRAY (1983), 'Mellor's Emeritus Headache', *Ratio*, 25: 81–8.*

—— (1986), 'Clipping Time's Wings', *Mind*, 95: 233–7.

—— (1988), 'Dummett's Second-Order Indexicals', *Mind*, 97: 113–16.

—— (1989), 'Omniscience and Eternity', *Aristotelian Society*, Supp. Vol. 63: 55–73.

McCALL, STORRS (1984), 'A Dynamical Model of Temporal Becoming', *Analysis*, 44: 172–6.

—— (1994), *A Model of the Universe* (Oxford: Clarendon Press).

McGINN, COLIN (1983), *The Subjective View* (Oxford: Clarendon Press).

MARKOSIAN, NED (1992), 'On Language and the Passage of Time', *Philosophical Studies*, 66: 1–26.

—— (1993), 'How Fast Does Time Fly?', *Philosophy and Phenomenological Research*, 53: 829–44.

MELLOR, D. H. (1980), 'The Self from Time to Time', *Analysis*, 40: 59–62.

—— (1981*a*), *Real Time* (Cambridge: Cambridge University Press).

—— (1981*b*), 'McTaggart, Fixity and Coming True', in Richard Healey (ed.), *Reduction, Time and Reality* (Cambridge: Cambridge University Press): 79–98.

MELLOR, D. H. (1986), 'Tense's Tenseless Truth-Conditions', *Analysis*, 46: 167–72.

—— (1989), 'I and Now', *Proceedings of the Aristotelian Society*, 89: 79–94.

—— (1995), *The Facts of Causation* (London: Routledge).

MOORE, A. W. (1987), 'Points of View', *Philosophical Quarterly*, 37: 1–20.

—— (1997), *Points of View* (Oxford: Clarendon Press).

NERLICH, GRAHAM (1994), *What Spacetime Explains: Metaphysical Essays on Space and Time* (Cambridge: Cambridge University Press).

NOONAN, HAROLD (1980), *Objects and Identity* (The Hague: Martinus Nijhoff).

—— (1985), 'A Note about Temporal Parts', *Analysis*, 45: 151–2.

—— (1988), 'Substance, Identity and Time', *Aristotelian Society*, Supp. Vol. 62: 79–100.

OAKLANDER, L. NATHAN (1984a), *Temporal Relations and Temporal Becoming: A Defense of a Russellian Theory of Time* (Lanham, Md.: University Press of America).

—— (1984b), 'McTaggart, Schlesinger, and the Two-Dimensional Time Hypothesis', *Philosophical Quarterly*, 33: 391–7.*

—— (1987), 'McTaggart's Paradox and the Infinite Regress of Temporal Attributions: A Reply to Smith', *Southern Journal of Philosophy*, 25: 425–31.*

—— (1990), 'The New Tenseless Theory of Time: A Reply to Smith', *Philosophical Studies*, 58: 287–92.*

—— (1991), 'A Defence of the New Tenseless Theory of Time', *Philosophical Quarterly*, 41: 26–38.*

—— (1992a), 'Thank Goodness It's Over', *Philosophy*, 67: 256–8.*

—— (1992b), 'Zeilicovici on Temporal Becoming', *Philosophia*, 21: 329–34.*

—— (1993), 'On the Experience of Tenseless Time', *Journal of Philosophical Research*, 18: 159–66.*

—— (1994a), 'McTaggart's Paradox Revisited', in Oaklander and Smith (1994: 211–13).

—— (1994b), 'Bigelow, Possible Worlds and the Passage of Time', *Analysis*, 54: 159–66.

—— (1996), 'McTaggart's Paradox and Smith's Tensed Theory of Time', *Synthese*, 107: 205–21.

—— and SMITH, QUENTIN (1994) (eds.), *The New Theory of Time* (New Haven: Yale University Press).

ODDIE, GRAHAM (1990), 'Backwards Causation and the Permanence of the Past', *Synthese*, 85: 71–93.

PADGETT, ALAN (1993), *God, Eternity and the Nature of Time* (London: Macmillan).

PERCIVAL, PHILIP (1989), 'Indices of Truth and Temporal Propositions', *Philosophical Quarterly*, 39: 190–9.

PLECHA, JAMES L. (1984), 'Tenselessness and the Absolute Present', *Philosophy*, 59: 529–34.

PRIEST, GRAHAM (1986), 'Tense and Truth-Conditions', *Analysis*, 46: 162–6.

—— (1987), 'Tense, *Tense* and TENSE', *Analysis*, 47: 184–7.

RICHARD, MARK (1981), 'Temporalism and Externalism', *Philosophical Studies*, 39: 1–13.

—— (1982), 'Tense, Propositions, and Meaning', *Philosophical Studies*, 41: 337–51.

SALMON, NATHAN (1989), 'Tense and Singular Propositions', in Joseph Almog et al. (eds.), *Themes from Kaplan* (Oxford: Oxford University Press): 331–92.

SAUNDERS, SIMON (1996), 'Time, Quantum Mechanics and Tense', *Synthese*, 107: 19–53.

SCHLESINGER, GEORGE N. (1980), *Aspects of Time* (Indianapolis: Hackett Publishing Company).

—— (1982), 'How Time Flies', *Mind*, 91: 501–23.

—— (1991), 'E Pur Si Muove', *Philosophical Quarterly*, 41: 427–41.

—— (1993), 'A Short Defence of Temporal Transience', *Philosophical Quarterly*, 43: 359–61.

—— (1995), *Timely Topics* (London: Macmillan).

SEDDON, KEITH (1987), *Time: A Philosophical Treatment* (New York: Croom Helm).

SHORTER, J. M. (1984), 'The Reality of Time', *Philosophia*, 14: 321–39.

SIMONS, PETER M. (1987), *Parts: A Study in Ontology* (Oxford: Clarendon Press).

—— (1991), 'On Being Spread Out in Time: Temporal Parts and the Problem of Change', in Wolfgang Spohn et al. (eds.), *Existence and Explanation* (Dordrecht: Kluwer): 131–47.

SMART, J. J. C. (1980), 'Time and Becoming', in Peter van Inwagen (ed.), *Time and Cause: Essays Presented to Richard Taylor* (Dordrecht: D. Reidel): 3–15.

SMITH, QUENTIN (1986), 'The Infinite Regress of Temporal Attributions', *Southern Journal of Philosophy*, 24: 383–96.*

—— (1987a), 'Sentences about Time', *Philosophical Quarterly*, 37: 37–53.

—— (1987b), 'Problems with the New Tenseless Theory of Time', *Philosophical Studies*, 52: 371–92.*

—— (1988), 'The Phenomenology of A-Time', *Diálogos*, 52: 143–53.*

—— (1988–9), 'The Logical Structure of the Debate about McTaggart's Paradox', *Philosophy Research Archives*, 24: 371–9.*

—— (1990a), 'Temporal Indexicals', *Erkenntnis*, 32: 5–25.*

—— (1990b), 'The Co-reporting Theory of Tensed and Tenseless Sentences', *Philosophical Quarterly*, 40: 213–22.*

—— (1993), *Language and Time* (New York: Oxford University Press).

—— and OAKLANDER, L. NATHAN (1995), *Time, Change and Freedom: An Introduction to Metaphysics* (London: Routledge).

SPRIGGE, T. L. S. (1991), 'The Unreality of Time', *Proceedings of the Aristotelian Society*, 92: 1–19.

STEIN, H. (1991), 'On Relativity Theory and the Openness of the Future', *Philosophy of Science*, 58: 147–67.

STUMP, ELEONORE, and KRETZMANN, NORMAN (1981), 'Eternity', *Journal of Philosophy*, 78: 429–58.

SWINBURNE, RICHARD (1990), 'Tensed Facts', *American Philosophical Quarterly*, 27: 117–30.

TEICHMANN, ROGER (1991), 'Future Individuals', *Philosophical Quarterly*, 41: 194–211.

TEICHMANN, ROGER (1995), *The Concept of Time* (London: Macmillan).

THOMSON, JUDITH JARVIS (1983), 'Parthood and Identity across Time', *Journal of Philosophy*, 80: 201–20.

TICHÝ, PAVEL (1980), 'The Transiency of Truth', *Theoria*, 46: 165–82.

TOOLEY, MICHAEL (1997), *Time, Tense, and Causation* (Oxford: Clarendon Press).

VAN INWAGEN, PETER (1990), 'Four-Dimensional Objects', *Noûs*, 24: 245–55.

WILLIAMS, CLIFFORD (1992*a*), 'The Phenomenology of B-Time', *Southern Journal of Philosophy*, 30: 123–37.*

—— (1992*b*), 'The Date Analysis of Tensed Sentences', *Australasian Journal of Philosophy*, 70: 198–203.*

YOURGRAU, PALLE (1986), 'On Time and Actuality: The Dilemma of the Privileged Position', *British Journal for the Philosophy of Science*, 37: 405–17.

—— (1987), 'The Dead', *Journal of Philosophy*, 84: 84–101.

—— (1990) (ed.), *Demonstratives* (Oxford: Oxford University Press).

ZEILICOVICI, DAVID (1986), 'A (Dis)Solution of McTaggart's Paradox', *Ratio*, 28: 175–95.

—— (1989), 'Temporal Becoming Minus the Moving Now', *Noûs*, 23: 505–24.*

INDEX OF NAMES

Ackrill, J. L. 22
Adams, R. M. 105
Alston, William 257 n.
Anselm, St 26
Aquinas, St Thomas 26–8, 223–31, 240
Aristotle 21–4, 26, 28, 50, 185, 192, 224, 226
Armstrong, David 104–5
Augustine, St 24–6, 27, 28, 37, 213, 252
Austin, J. L. 63–4
Ayer, A. J. 81 n.

Balázs, Béla 277 n.
Barnes, A. 129
Benn, Piers 8
Bernstein, M. H. 185 n.
Boas, F. 70
Boethius 26, 51
Bohr, Niels 137
Broad, C. D. 19–20, 23, 24, 36, 46, 47, 48 n., 189–90, 198, 202–3
Butterfield, Jeremy 5, 62 n., 69

Cahn, Steven 185 n.
Callender, Craig 138 n.
Cantor, Georg 176
Carter, W. 186 n.
Chao, Y. 70 n.
Cicero 68 n.
Coburn, R. 221 n.
Cockburn, David 5
Craig, William Lane 9, 28 n., 136 n., 142, 174, 241, 247
Cresswell, Max 101–3
Currie, Gregory 10, 115 n., 270 n., 271 n., 275 n., 276 n., 278 n., 279 n., 282 n.

Davidson, Donald 29
Davies, Martin 100–1
Davis, S. 223 n., 235 n.
Denbigh, K. 246 n.
Dummett, Michael 4, 17, 61, 62, 69
Dyke, Heather 6, 93 n., 94 n.

Einstein, Albert 135, 138–40, 144, 245
Epicurus 209, 210, 216

Farmer, David J. 17
Faulconer, B. 62 n.
Feinberg, G. 129
Findlay, J. N. 26
Flowers, J. 62 n.
Forbes, Graeme 99–100, 145
Friedman, Michael 129

Gale, R. M. 14, 29, 61, 221 n.
Garson, J. 62 n.
Geach, Peter 186 n., 228–9, 230
Genette, Gérard 278 n.
Goldhaber, A. 129
Goodman, Nelson 28, 29
Grünbaum, Adolf 244–5

Harris, J. 223 n.
Hasker, W. 235 n.
Heisenberg, Werner 137
Heller, M. 186 n., 201 n.
Helm, Paul 9, 28 n., 221 n., 235 n., 248, 262 n.
Hestevold, H. Scott 186 n.
Hill, W. 229
Hoerl, Christoph 39–40
Hume, David 52 n.

Isham, Christopher 140

Jackson, Frank 274
Jantzen, G. 221 n.

Kretzmann, Norman 9, 28 n., 231–40, 245, 248
Kripke, Saul 274 n.

LaBerge, D. 62 n.
La Croix, R. 223 n.
Leftow, Brian 174 n., 221 n., 235 n., 239, 240–6
Leibniz, G. W. 104, 140
Leith, John 251 n.
Le Poidevin, Robin 6, 16, 20, 37, 38 n., 49 n., 55 n., 80, 94 n., 102 n., 109 n., 112–13, 115, 116, 186 n., 192 n., 198 n., 247 n., 262, 263, 265, 270 n., 271–4

Levison, A. B. 38
Lewis, David 52 n., 53, 93, 94–7, 98, 101,
  103–9, 113–14, 116, 194 n., 202,
  272 n., 273–6, 279
Lewis, Delmas 235 n.
Liske, M.-T. 223
Lloyd, Genevieve 61
Lombard, Lawrence 186 n.
Lorentz, Hendrik 141–4, 145, 171
Lotman, J. 277 n.
Lowe, E. J. 3, 32–3, 34, 35, 39, 47, 49,
  53 n., 98–9, 109, 110–12, 115, 116
Lucas, J. R. 185 n., 186–7, 221 n.
Łukasiewicz, J. 186 n.
Lycan, William 106–8, 113

MacBeath, Murray 83 n., 187 n., 248 n.
McCall, Storrs 43 n., 48 n., 195
McCann, H. 228 n.
McTaggart, J. M. E. 4, 13–20, 23, 26, 27,
  28, 31, 36, 37 n., 38, 44, 46–7, 101–3,
  110, 115, 185, 190 n., 210, 216, 218,
  236, 265
Mann, W. 221 n.
Markosian, Ned 187 n.
Maxwell, James Clerk 142–3
Mellor, D. H. 3, 6, 20, 28, 30, 31–2, 33, 34,
  43 n., 45 n., 61, 62, 81 n., 87, 88–9,
  93, 101–3, 132 n., 187 n., 192 n.,
  198 n., 216, 248
Mendilow, A. A. 277
Mignani, R. 129
Miller, Richard B. 107–8, 113
Mitchell, R. 62 n.

Nagel, Thomas 81–5, 89–90, 207
Naylor, Margery Bedford 106
Nerlich, Graham 7, 128 n., 129, 130 n.,
  140–1, 157, 163 n., 177
Newton, Isaac 138 n., 168, 173, 233
Newton-Smith, W. H. 10, 172
Nieto, M. 129

Oaklander, L. Nathan 8, 10 n., 29, 37, 39,
  159–60, 186 n., 187 n., 190 n., 193,
  198, 201 n.
Olli, J. 70 n.
Owens, David 143

Padgett, Alan 28 n., 188 n., 190–1, 193–4,
  221 n., 242 n., 248 n.
Parfit, Derek 3–4, 8
Penrose, Roger 141 n.
Perrett, Roy 209

Perry, John 28
Pike, Nelson 221 n.
Plantinga, Alvin 145
Posner, M. 62 n.
Potter, M. 62 n.
Priest, Graham 32
Prior, A. N. 5, 21, 24, 26, 38, 49, 61, 82–3,
  112
Putnam, Hilary 7, 61

Quine, W. V. O. 61, 69 n., 70, 135, 138,
  147

Ramsey, F. P. 120
Recami, E. 129
Reichenbach, Hans 39
Ricœur, Paul 271 n.
Robb, A. A. 129, 163 n.
Robinson, Abraham 179 n.
Russell, Bertrand 5, 13, 16–18, 21, 26, 27,
  28, 29, 198

Salmon, Nathan 106
Schlegel, Richard 137
Searle, John 120
Seddon, Keith 85 n., 246 n.
Shaffer, W. 62 n.
Shanks, Niall 185 n., 186 n., 195–201
Sidelle, Alan 169–70
Silverstein, Harry 209–10
Sklar, Lawrence 245
Smart, J. J. C. 29
Smith, Quentin 7, 10 n., 29–30, 34–5, 36,
  38, 39, 46, 47, 48, 49 n., 109–10, 111,
  115, 116, 119–29, 133, 136, 138, 139,
  153, 157, 162 n., 164 n., 165, 167–8,
  169–70, 171, 174 n., 175 n., 186 n.,
  187 n., 188 n., 190 n., 195 n., 228 n.
Sorabji, Richard 24
Spinoza, Benedict de 5, 77–80, 81–2, 90,
  91
Stalnaker, Robert 105
Stein, Howard 195 n.
Stump, Eleonore 231–40, 245, 248
Swinburne, Richard 9, 167–8, 253–62

Talib, I. S. 280–1
Taylor, Richard 61, 62
Tooley, Michael 2 n., 33 n., 36, 37, 39,
  43 n., 46 n., 147, 192 n., 195 n.,
  254 n.

Vallicella, William F. 153–4, 156, 157–8
van Inwagen, Peter 101

Walker, Ralph 221 n.
Weingard, Robert 138 n.
Wells, H. G. 272 n.
Wierenga, E. 223 n.
Williams, Bernard 67 n.
Williams, Clifford 187 n., 198, 199 n.
Williams, D. C. 186 n.
Wilson, N. L. 68–9
Wittgenstein, Ludwig 90
Wolff, Robert Paul 275 n.

Wolterstorff, Nicholas 228 n.
Wright, John 231

Yagisawa, Takashi 106
Yates, J. 221 n., 223 n., 228 n., 235 n.
Yourgrau, Palle 98, 185 n., 186, 211 n.

Zahar, Eli 142, 143–4
Zemach, Eddy 62